CLINICAL CHEMISTRY HANDBOOK

WORKBOOK OF PRINCIPLES, TECHNIQUES AND CORRELATIONS

N. T. COLEMAN

outskirts press

Clinical Chemistry Handbook
Workbook of Principles, Techniques and Correlations
All Rights Reserved.
Copyright © 2022 N. T. Coleman
V3.0

The opinions expressed in this manuscript are solely the opinions of the author and do not represent the opinions or thoughts of the publisher. The author has represented and warranted full ownership and/or legal right to publish all the materials in this book.

This book may not be reproduced, transmitted, or stored in whole or in part by any means, including graphic, electronic, or mechanical without the express written consent of the publisher except in the case of brief quotations embodied in critical articles and reviews.

Outskirts Press, Inc.
http://www.outskirtspress.com

ISBN: 978-1-9772-3768-2

Library of Congress Control Number: 2022906257

Cover Photo © 2022 www.gettyimages.com. All rights reserved - used with permission.

Outskirts Press and the "OP" logo are trademarks belonging to Outskirts Press, Inc.

PRINTED IN THE UNITED STATES OF AMERICA

Contents

CHAPTER 1: Photometry 1
CHAPTER 2: Electrochemistry 7
CHAPTER 3: Chromatography 10
CHAPTER 4: Mass Spectrometry 14
CHAPTER 5: Electrophoresis 18
CHAPTER 6: Immunochemical Techniques 21
CHAPTER 7: Analytical and Clinical Evaluation of Methods 25
CHAPTER 8: Molecular Diagnostics 32
CHAPTER 9: Molecular Techniques 38
CHAPTER 10: Therapeutic Drug Monitoring (TDM) 42
CHAPTER 11: Amino Acid and Proteins 50
CHAPTER 12: Non-Protein Nitrogen Compounds 59
CHAPTER 13: Enzymes 68
CHAPTER 14: Carbohydrate 77
CHAPTER 15: Lipids and Lipoproteins 85
CHAPTER 16: Electrolytes and Blood Gases 89
Index 95
Reference 96

Preface

We are pleased to introduce the first edition of Clinical Chemistry Handbook a Workbook of Principles, Techniques and Correlations. We built on the excellent work of our predecessors and used electronic tools to produce a state-of-the-art product for students, trainees, and practicing clinical laboratory scientists. We aimed to harmonize the presentation of information among chapters while focusing on the clinical significances of each aspect of clinical chemistry. Unlike most other workbooks, all chapters in this edition were reviewed by three individuals: a reviewer, an associate editor, and a senior editor. We believe that these efforts have led to a better product. In addition, we made a concerted effort to create an international rather than an American product to reflect different practices from around the world; for example, all measurements are presented both in traditional and SI units and included the glossary within the chapters rather than at the end of the book. We sincerely hope that this product will be a valuable educational and reference resource for the clinical laboratory scientists' community worldwide.

N. T. Coleman

CHAPTER 1
Photometry

Clinical Significance

Many determinations made in the clinical laboratory are based on measurements of radiant energy interacting with body fluids. Photometry methods is established on measurements of light by a photodetector. If light is absorbed by a substance dissolved in solution an absorbance measurement is obtained. Light may also be scattered or refracted by particles suspended in solution and this light is detected using a turbidimetry or nephelometry instrument. Light may be emitted from a substance that absorbs light at one wavelength and emits light at another wavelength; this is fluoresces spectrometry. Specific wavelengths of light are chosen for each analysis based on the properties of the substance being measured. A typical light source (lamp) generates a broad range of wavelengths of light. A visible lamp produces light of wavelengths from 400 nm (violet light) to 700 nm (red light). An ultraviolet lamp produces light of wavelengths from about 200 to 400 nm. To select the desired wavelength from the spectrum of light produced by the light source, a device called a monochromator or filters are used. A monochromator disperses the light (much like a prism disperses light) and allows selection of a narrow band of wavelengths to be directed through the sample cuvette (1).

The type of radiant energy
- Emitted
- Transmitted
- Absorbed
- Scattered
- Reflected under controlled conditions

Definitions and Key Terms

Absorbance (Abs)
The amount of light absorbed as incident light passes through a sample, which is equivalent to log (1/T), where T is transmittance. Absorbance and transmittance are inverse properties.

Absorptivity (a)
A proportionality constant for a compound that is the measure of the absorption of energy at a given wavelength as it passes through a solution of that compound at a concentration of 1 g/L; expressed mathematically as absorbance divided by the product of the concentration of a substance in g/L and the sample path length in centimeters ($a = Abs/bc$).

Atomic absorption spectrophotometry
A technique in which an element in a sample is dissociated from its chemical bonds (atomized) and placed in an unexcited or ground state (neutral atom); the atom in the ground state is able to absorb radiation and the decrease in radiant energy from the light source transmitted through the sample is measured.

Beers law
A mathematical equation stating that the concentration of a substance is directly proportional to the amount of light absorbed, mathematically expressed as $Abs = abc$ or $Abs = \varepsilon bc$.

Chemiluminescence
The emission of light when a molecule returns from an excited or higher energy level to a lower energy level, in which the excitation event is caused by a chemical reaction and not by photo illumination, for example, by the oxidation of an organic compound.

Electrochemiluminescence
The emission of light when a molecule returns from an excited or higher energy level to a lower energy level, in which the excitation event is a reaction generated electrochemically on the surface of an electrode.

Flow cytometry
The measurement of optical properties of cells or particles made while they pass singly through a measuring apparatus in a flowing fluid stream.

Fluorescence
The emission of electromagnetic radiation that occurs when a molecule absorbs light at one wavelength and reemits light at a longer wavelength.

Fluorometry
The measurement of emitted fluorescence light that occurs when a molecule absorbs light at one wavelength and reemits light at a longer wavelength.

Light
The energy transmitted via electromagnetic waves that are characterized by frequency and wavelength; light is composed of photons whose energy is inversely proportional to the wavelength.

Light scattering
The redirection of light that results from the interaction of light with molecules or particles in solution without absorption taking place.

Molar absorptivity (ε)
A proportionality constant for a compound that is the measure of the absorption of radiant energy at a given wavelength as it passes through a solution of that compound at a concentration of 1 mol/L; expressed mathematically as absorbance divided by the product of the concentration of a substance in mol/L and the sample path length in centimeters ($\varepsilon = Abs/bc$).

Nephelometry
The detection and measurement of light energy scattered or reflected toward a detector that is not in the direct path of the transmitted light; nephelometers may measure scattered light at right angles to the incident light, or at some angle other than 90 degrees.

Spectrophotometry
The measurement of the luminous intensity of light or the amount of luminous light falling on a surface of a detector; spectrophotometry is the measurement of the intensity of light at varied wavelengths.

Reflectance photometry
A spectrophotometric technique in which diffused light illuminates a reaction mixture in a carrier containing a substance of interest, and the intensity of the reflected light is measured and compared with a reference.

Spectral bandwidth
The width in nanometers of the spectral transmittance curve at a point equal to one-half of the peak transmittance; used to describe the spectral purity of a filter or other monochromator.

Transmittance
The intensity of a light beam that passes through a square cell containing a solution of a compound that absorbs light at a specific wavelength, divided by the intensity of an incident (incoming) light beam, stated as $T = I/I_0$.

Turbidimetry
The detection and measurement of a decrease in intensity of an incident beam of light after it passes through a solution of molecules or particles.

Wavelength (λ)
A characteristic of electromagnetic radiation; the distance between two adjacent maxima in a wave that is measured in nanometers.

Photometry Problems

1. Describe the relationship between the light transmitted and the light absorbed by a solution of a compound; state this relationship as a mathematical formula.

2. Express Beer's law mathematically, and define each of the components of the formula; calculate the concentration of a substance in solution using Beer's law.

3. Explain how Beer's law is used to create a calibration curve, and list five conditions that must be met before Beer's law can be applied to a measurement.

4. List the basic components of a single-beam and double-beam spectrophotometer and state the purpose of each of the components; provide and describe two examples, of each component.

5. State the principle of reflectance photometry and the clinical laboratory applications for this technique

6. What is the principle of Atomic Absorbance (AA) spectrophotometry? List the clinical laboratory applications.

7. What are the principles of fluorescence spectrophotometry? List clinical laboratory applications

8. Compare the sensitivity of fluorescence intensity measurements with that of absorbance measurements, and list reasons why fluorescence measurements are more sensitive.

9. Compare nephelometry and turbidimetry, and list the clinical laboratory applications of these techniques.

10. Many determinations made in the clinical laboratory are based on measurements of radiant energy (1) emitted, (2) transmitted, (3) absorbed, (4) scattered, or (5) reflected. What instrument is used for each type of radiant energy listed?

11. What is the concentration of serum uric acid in a solution whose absorbance is 0.098, cuvet pathlength, 1.0 centimeters, absorptivity is 0.102 L $g^{-1}cm^{-1}$?
    a. 0.96 mg/L
    b. 0.009 g/L
    c. 0.48 g/L
    d. 1.96 g/L
    Answer: a

12. Which of the following types of light sources emits radiation that changes in intensity very slowly as a function of wavelength?
    a. Pulsating source
    b. Line source
    c. Continuum source
    d. Microwave source
    Answer: c

Photometry

13. Filters and diffraction gratings are examples of which of the following?
    a. Monochromators
    b. Cuvettes
    c. Light sources
    d. Slits
    Answer: a

14. The wavelength in nanometers at peak transmittance is termed:
    a. Spectral bandwidth.
    b. Stray light.
    c. Nominal wavelength.
    d. Electromagnetic spectrum.
    Answer: c

15. Bandpass is defined as:
    a. The range of wavelengths at a point halfway between the baseline and the peak.
    b. The half power point or full width at half maximum.
    c. The total range of wavelengths transmitted.
    d. The product of parallel beams of radiation.
    Answer: c

16. A photomultiplier tube function is to:
    a. Reduce the electronic signal produced as photons cascade from dynode to dynode.
    b. Increase the electronic signal produced as photons cascade from dynode to dynode.
    c. Decrease the number of transmitted photons from the cuvette.
    d. Reduce in a logarithmic fashion the electronic signal created at the cathode.
    Answer: b

17. Holmium oxide along with Didymium filters are used to assess:
    a. Wavelength accuracy.
    b. Voltage values.
    c. Stray light.
    d. Photometric accuracy.
    Answer: a

18. Which of the following describes atomic absorption spectroscopy?
    a. The transmittance of monochromatic electromagnetic radiation by an element in its excited state
    b. The measurement of the electromagnetic radiation emitted by an element
    c. The release of electromagnetic radiation due to a chemical reaction
    d. The absorption of monochromatic electromagnetic radiation by an element in its ground state with a net zero charge
    Answer: d

19. Which of the following statements best illustrates the difference between photons from fluorescent emission and the excitation photons?
    a. Emission photons are higher energy and longer wavelength.
    b. Emission photons are lower energy and longer wavelength.
    c. Emission photons are lower energy and shorter wavelength.
    d. Emission photons are negatively charged and have zero mass.
    e. Answer: b

20. The polarization of light in fluorescent polarization immunoassay allows the measurement of which of the following?
    a. Bound fraction in the presence of the free fraction
    b. Only the bound fraction after the free fraction is removed
    c. Both bound and free fractions
    d. Antibodies to fluorescent label
    Answer: a

21. A chelate of europium (Eu^{3+}) is used in which of the following assays?
    a. Radioimmunoassays
    b. Chemiluminescent immunoassays
    c. Time-resolved fluorescent immunoassays
    d. Fluorescent polarization immunoassays
    Answer: c

22. Oxidation of acridinium esters by hydrogen peroxide in the presence of peroxidase produces which of the following?
    a. Fluorescence
    b. Phosphorescence
    c. Translucence
    d. Chemiluminescence
    Answer: d

23. Turbidimetry is the measurement of the reduction in light transmission caused by:
    a. Particle formation.
    b. Chelation of metals.
    c. Slowing of the spin rotation of electrons.
    d. Increase in excitation photon energy.
    Answer: a

24. The presence of particulates in a urine sample measured by a refractometer will result in which of the following?
    a. An increase in reflectance of the solution
    b. A change in the critical angle of the light as it passes through the solution
    c. An increase in energy of the solute particles that results in a release of light energy
    d. A change in Rayleigh scatter patterns
    Answer: b

CHAPTER 2
Electrochemistry

Clinical Significance

Electrodes are electrochemical probes that measure ions in solution. They are used in clinical chemistry is to measure proton (H^+) concentration using a pH meter. Ion-selective electrodes selectively measure one specific ion in the presence of other ions. The technology of ion-selective electrodes is used in many clinical instruments and is particularly useful for measuring electrolytes(2, 3).

Definitions and Key Terms

Activity of an ion
The concentration of free, unbound ion in solution.

Amperometry
An electrolytic electrochemical process in which current is monitored at a fixed voltage between working and reference electrodes in an electrochemical cell.

Biosensor
A type of chemical sensor consisting of a biologic recognition element and a physicochemical transducer, often an electrochemical or an optical device.

Conductometry
An electrochemical technique used to determine the quantity of an analyte present in a mixture by measuring its effect on the electrical conductivity of the mixture.

Coulometry
An electrochemical technique that measures the electrical charge passing between two electrodes in an electrochemical cell, with the amount of charge passing between the electrodes being directly proportional to oxidation or reduction of an electroactive substance at one of the electrodes.

Electrochemical cell
A device that consists of two electrodes (electron or metallic conductors) connected by an electrolyte solution that conducts ions (galvanic cell), or a device in which an external voltage is applied to a polarizable working electrode versus a reference electrode, with the resulting cathodic or anodic current of the cell being monitored (electrolytic cell).

Electrode
A half-cell that consists of a single metallic conductor in contact with an electrolyte solution; the indicator (measuring) electrode is one half-cell and the reference electrode is the second half-cell.

Electrode potential
The electromotive force (EMF) of a single half-cell measured with respect to the standard hydrogen electrode, set at zero by convention.

Ion-selective electrode
An electrode that selectively interacts with a single ionic species; the potential produced at the membrane/sample solution interface is proportional to the logarithm of the ionic activity or the concentration of the ion in question.

Nernst equation
The equation used to relate the potential of an electrochemical cell to the activity of a chemical species in solution.

Optode
An optical sensor that measures specific substances such as pH, blood gases, and electrolytes, using dye immobilization, fluorescence quenching, or phosphorescence.

Potential difference
The work required to move an electrical charge and measured in volts.

Potentiometry
An electrochemical technique that measures an electrical potential difference between two electrodes (half-cells) in an electrochemical cell.

Electrochemistry Problems

1. What are the clinical applications for each of the following electrochemical measurement techniques: (a) potentiometry, (b) voltammetry/amperometry, (c) conductometry, and (d) coulometry.

2. List and describe the components of a basic ion selective electrode

3. Define the following: (a) amperometry, (b) biosensor, (c) conductometry, (d) coulometry, (e) electromotive force, (f) indicator electrode, (g) ion activity, (h) molality, (i) optode, (j) potentiometry, (k) reference electrode, and (l) voltammetry.

4. What are the two major categories of biosensors and give a practical example of each type in the clinical laboratory?

5. An optical sensor used in analytical instruments to measure pH, blood gases, and electrolytes is referred to as a(n):

6. A voltmeter that measures the potential across an electrochemical cell (between the two electrodes) is referred to as a:

7. An optical sensor used in analytical instruments to measure pH, blood gases, and electrolytes is referred to as a(n):

CHAPTER 3
Chromatography

Clinical Significance

Clinical use of chromatography can include drug analysis, therapeutic drug monitoring (TDM) and is used to understand the nutrient status of patients. While urine is the most common medium for analyzing biofluid concentrations, blood serum and plasma are collected for most medical analyses with chromatography instruments.

Definitions and Key Terms

Adsorption chromatography
A type of liquid chromatography in which chemicals are retained based on their adsorption and desorption at the surface of a non-derivatized support.

Affinity chromatography
A liquid chromatographic method that makes use of intermolecular forces (IMF) interactions for the retention and separation of chemicals.

Anion-exchange chromatography
A type of ion-exchange chromatography that uses positively charged groups as a stationary phase to bind and separate anions.

Bonded phase gas chromatography
A type of gas chromatography in which the stationary phase is chemically linked or attached to the support.

Carrier gas
A term often used to describe the mobile phase in gas chromatography.

Cation-exchange chromatography
A type of ion-exchange chromatography that uses negatively charged groups as a stationary phase to bind and separate cations.

Chiral stationary phase
A stationary phase that is chiral and that can interact with compounds in a stereospecific manner.

Chromatogram
A plot of detector response in chromatography as a function of the time or volume of the mobile phase that is needed to elute an analyte from a column.

Chromatograph
The instrument that is used to perform a separation in chromatography.

Chromatography
A method in which the components of a mixture are separated based on their differential interactions with two chemical or physical phases: a mobile phase and a stationary phase.

Column chromatography
A type of chromatography that uses a column, or a tube, to contain the stationary phase and support.

Gas chromatography (GC)
A type of chromatography that uses a gas as the mobile phase.

Gas chromatography/mass spectrometry (GC/MS)
The use of a mass spectrometer as a detector for gas chromatography.

Gas-liquid chromatography (GLC)
A type of gas chromatography in which the stationary phase is a liquid that is placed as a coating or layer on the support.

Gas-solid chromatography (GSC)
A type of gas chromatography in which the same material acts as both the stationary phase and the support; chemicals are retained by their adsorption to the surface of the support.

High-performance liquid chromatography (HPLC)
A type of liquid chromatography that uses small, efficient supports that produce narrow peaks.

Hydrophilic interaction liquid chromatography (HILIC)
A type of partition chromatography that uses a polar stationary phase, and in which chemicals partition between an organic-rich region in the mobile phase and a more polar water-enriched layer that is at or near the surface of a polar support.

Internal standard
A compound that is added in a constant amount to standard solutions and samples to help normalize the results for any variations that may occur during pretreatment steps or during injection into a chromatographic system.

Ion-exchange chromatography
A type of liquid chromatography in which ions are separated by their adsorption onto a support that contains fixed charges at its surface.

Liquid chromatography–electrochemical detection (LC-EC)
The combination of an electrochemical detector with liquid chromatography.

Liquid chromatography–mass spectrometry (LC-MS)
A technique that combines liquid chromatography with mass spectrometry.

Mobile phase
The phase that travels through a chromatographic system and carries sample components with it.

Normal-phase chromatography
A type of partition chromatography that uses a polar stationary phase; also known as "normal-phase liquid chromatography.

Stationary phase
The phase in chromatography that is held within the system by a support and used to interact with and separate a sample's components as these components pass through the system.

Reversed-phase chromatography
A type of partition chromatography that uses a nonpolar stationary phase.

Chromatography Problems

1. Explain how external or internal calibration can be used for quantitative measurements in chromatography.

2. Define internal standard and discuss how this can be used in quantitative measurements by chromatography.

3. Define gas chromatography (GC) and describe three types of GC based on the type of stationary phase that is present.

4. How is liquid chromatography and high-performance liquid chromatography different?

5. List several ways in which liquid chromatography differs from GC.

6. Which component in a chromatographic system carries the sample?

7. Describe the following types of liquid chromatography, including the principles of separation, the types of stationary phases and mobile phases that are used, and some possible clinical applications of each.
    a) Adsorption chromatography
    b) Affinity chromatography
    c) Hydrophilic interaction liquid chromatography
    d) Ion-exchange chromatography
    e) Ion-pair chromatography
    f) Normal-phase chromatography
    g) Reversed-phase chromatography

CHAPTER 4
Mass Spectrometry

Clinical Significance

Mass spectrometers coupled with gas and liquid chromatographs (GC-MS and LC-MS) are versatile analytical instruments that allow high specificity, high sensitivity, and high throughput assay development of complex samples. Gas chromatography–mass spectrometry (GC-MS) has been used formally years in the analysis of biological samples. analytes. One of the most common applications of GC-MS is in drug testing for clinical or forensic purposes. Because many drugs have relatively low molecular weight and are sufficiently volatile, they are suitable for analysis by GC. Identification of microorganisms using MALDI-TOF has become established for routine use in clinical laboratories (1, 4).

Some examples of the clinical use of GC-MS methods other than drug testing are in the quantitative analysis of:
- Xenobiotics
- Steroids
- Pesticides
- Pollutants
- Metabolic inborn errors

LC-MS/MS methods are widely used in the areas of:
- Newborn screening for metabolic disorders
- Endocrinology
- Toxicological drug screening
- Therapeutic drug monitoring
- Cancer biomarkers
- Analysis of biomarker

In clinical laboratories, ICP-MS is used for quantification of trace elements in whole blood, urine, plasma, serum, and tissue biopsy samples. Clinical reasons for analyzing biological samples for trace elements include testing for suspected heavy metal poisoning, and diagnosing/monitoring patients with metabolic disorders (e.g., Wilson disease) (5).

Definitions and Key Terms

Base peak
An ion with the highest abundance in the mass spectrum; it is assigned a relative abundance of 100%.

Electrospray ionization
Using this ionization technique, a sample is passed through a capillary to which a high voltage is applied, resulting in expulsion of ions from a spray of fine droplets.

Extracted ion chromatogram
A plot consisting of a sum of selected mass-to-charge ratios displayed as a function of time.

Fragment ion
An ion formed by fragmentation of a molecular ion.

Ion
An atom or a molecule that has acquired an electrical charge by losing or gaining one or more electrons or other charged species such as protons.

Ionization
Required for all mass spectrometry techniques; "ionization" describes the production of ions from neutral atoms or molecules using various techniques including but not limited to electron ionization, chemical ionization, or electrospray ionization.

Ion trap
Type of mass analyzer in which ions are held in a spatially confined region using, for example, a magnetic or electrostatic field. Manipulation of the trap allows m/z measurements to be performed, as well as fragmentation of the molecular ion to allow multiple levels of the fragmentation (MSn).

Isotope
A variant of a chemical element; each variant differs in the number of neutrons in the nucleus and therefore in atomic weight.

Isotope dilution mass spectrometry (IDMS)
An analytical technique used to quantify a compound relative to isotopic analogs added to samples as internal standards at fixed concentrations.

Liquid chromatography–mass spectrometry (LC-MS)
A technique that uses a liquid chromatograph coupled to a mass spectrometer. Mass-to-charge ratio (m/z)
The ratio formed by dividing the molecular mass of an ion by its charge.

Mass analysis
The process of resolving ionic species based on their mass-to-charge ratios.

Mass spectrometer
An analytical instrument that ionizes atoms or molecules; separates the ions and detects the m/z of these ionized atoms or molecules or their ion fragments. Mass spectrometers are often interfaced with other

instruments including but not limited to a second mass spectrometer, a liquid chromatograph, or a gas chromatograph.

Mass spectrometry (MS)
Study of matter through the formation of gas-phase ions that are characterized by their mass, charge, structure, and/or physicochemical properties.

Mass spectrum
A plot of the relative abundance of ions plotted as a function of their mass-to-charge ratio.

Matrix-assisted laser desorption/ionization (MALDI)
A soft ionization technique allowing analysis of organic molecules (proteins, peptides, DNA, polysaccharides), and microorganisms. MALDI ionization is performed by irradiation of sample mixed with a matrix using pulsed laser.

Product ion
A fragment ion formed when a molecular ion breaks into smaller pieces. In a tandem mass spectrometer, the fragmentation process takes place after molecular ions have been separated by their m/z value in the first stage.

Selected ion monitoring (SIM)
A mass spectrometry technique where signal from only specified ions of interest are acquired.

Tandem mass spectrometer
A mass spectrometer capable of multistage separation of ions in an ion beam based on the m/z values (tandem in space) or separation of trapped ions according to m/z values (tandem in time).

Total ion chromatogram (TIC)
A chromatogram created by summing up intensities of all m/z responses acquired by a mass analyzer during chromatographic separation. The chromatogram is displayed as a function of time.

Mass Spectrometry Problems

1. For the following ionization methods, state the principle and specific uses of each type:
    a. Electron impact
    b. Chemical ionization
    c. Electrospray
    d. Atmospheric pressure chemical
    e. Inductively coupled plasma (ICP)
    f. MALDI

2. State the principles of operation of quadrupole, triple quadrupole, ion trap, and TOF mass analyzers.

3. List types of commonly used electron multiplier detectors.

4. Define the following terms:
    a. Base peak
    b. Ion trap
    c. Mass analysis
    d. Mass spectrometry
    e. Mass-to-charge ratio
    f. Molecular ion
    g. Product ion
    h. Fragment ion

5. In mass spectrometry, the sum of all ions produced displayed as a function of time represents:
    a. product ion
    b. extracted ion profile
    c. mass spectrum.
    d. total ion chromatogram
       Answer d

CHAPTER 5
Electrophoresis

Clinical Significance

Protein electrophoresis is an electrophoretic method of separating proteins present in the serum or plasma, based on their molecular weight and electric charges. Electrophoresis had been extensively used in clinical medicine for aiding in diagnosis of various clinical conditions like acute and chronic inflammations, monoclonal gammopathies, nephropathy, liver diseases and cardiac injury (6-9).

Definitions and Key Terms

Ampholyte
A molecule that is positively or negatively charged on the basis of the pH of the solution in which it resides; proteins, because they contain many ionizable amino and carboxyl groups, behave as ampholytes in solution and are considered amphoteric.

Capillary electrophoresis
A method in which the classic techniques of electrophoresis are carried out in a small-bore, fused silica capillary tube coated with a polymeric covering.

Densitometry
A measuring technique that uses an optical system to scan and quantify electrophoretic fractions separated on a gel or other medium.

Electropherogram
A densitometric display of protein zones on a support material after separation and staining.

Electrophoresis
The migration of charged solutes or particles within a liquid medium under the influence of an electrical field.

Electrophoretic mobility (μ)
The rate of migration (cm/s) of a charged solute in an electrical field, expressed per unit field strength (volts/cm).

Endosmosis (electroendosmotic flow)
Preferential movement of water in one direction through an electrophoresis medium due to selective binding of one type of charge on the surface of the medium.

Isoelectric focusing
An electrophoretic technique that separates amphoteric compounds within a medium that possesses a stable pH gradient.

Isoelectric point (pI)
The pH at which a molecule has no net charge and will not migrate during electrophoresis.

Wick flow
Movement of water from the buffer reservoirs toward the center of an electrophoresis gel or strip to replace water lost by evaporation.

Zone electrophoresis
Migration of charged molecules in an applied electric field.

Electrophoresis Problems

1. What is the relationship between pH and pI?

2. What does it mean for a protein to be amphoteric?

3. How does protein charge effect migration?

4. What optical measuring technique allows for the scanning and quantification of electrophoretic fractions separated on a support medium?

5. Describe how the following affect electrophoresis:
    a. size and shape of molecule.
    b. heat and current produced by the power supply.
    c. buffer pH, and type of gel used.

6. Describe and compare the following types of electrophoretic techniques, including clinical utility, system components, and detection methods:
    a. Capillary electrophoresis
    b. isoelectric focusing electrophoresis
    c. microchip electrophoresis
    d. two-dimensional electrophoresis
    e. zone electrophoresis

7. What are the advantages of capillary over conventional electrophoresis?

8. 5) Which of the following is TRUE about gel electrophoresis of DNA?
    a. DNA of smaller molecular weight will travel far from the point of application.
    b. DNA migrates toward the cathode.
    c. Increasing the voltage increases the resolution of DNA fragments separated in the gel.
    d. DNA electrophoresis requires polyacrylamide as a matrix.
    Answer: c

CHAPTER 6
Immunochemical Techniques

Clinical Significance

Immunochemistry offers simple, rapid, robust yet sensitive, and in most cases, easily automated methods applicable to routine analyses in clinical laboratories. Immunochemical methods have rapidly replaced chromatographic techniques in clinical diagnostics, offering fast detection of antibodies associated with specific diseases, disease biomarkers, hormones, and pharmaceuticals. The assays most often used in clinical immunochemistry involve either quantitative or qualitative formats using enzyme- linked immunosorbent assays (ELISAs), immunochromatography in the form of lateral-flow devices like dip-sticks and test strips or Western Blot assays used to interpret data from protein analysis with gel electrophoresis. Similarly, immunohistochemistry, one of the main diagnostics tools in today's clinical laboratories, is also based on the principles of antigen-antibody binding (10, 11).

Definitions and Key Terms

Affinity
Energy of interaction of a single antibody-combining site and its corresponding epitope on the antigen.

Antibody
Immunoglobulin (Ig) class of molecule (e.g., IgA, IgG, IgM) that binds specifically to an antigen or hapten.

Antigen
Any material capable of reacting with an antibody without necessarily being capable of inducing antibody formation.

Avidity
Overall strength of binding of antibody and antigen; includes the sum of the binding affinities of all individual combining sites on the antibody.

Competitive immunoassay
An immunoassay in which all reactants are simultaneously or sequentially mixed together and unlabeled antigen competes with labeled antigen for binding sites on the antibody.

Hapten
A chemically defined factor that, when conjugated to an immunogenic carrier, stimulates the synthesis of antibody specific for the hapten.

Heterogeneous immunoassay
An immunochemical reaction in which it is assumed that the formation of the antigen/antibody complex occurs more quickly than the breakdown of the complex into antigen and antibody; in this assay, the antigen is labeled and separation of the free from the bound labeled antigen is required.

Homogeneous immunoassay
An immunochemical reaction in which the activity of the label attached to the antigen is modulated directly by antibody binding; this assay does not require a separation.

Hook effect
A phenomenon occurring with certain immunoassays due to very high concentrations of a particular analyte; it results in a false-negative result. The hook effect mostly affects one-step immunometric assays.

Immunoassay
An assay based on the reaction of an antigen with an antibody specific for the antigen.

Immunogen
A substance capable of inducing an immune response.

Label
Any substance with a measurable property attached to an antigen, antibody, or binding substance.

Noncompetitive immunoassay
An immunoassay in which a capture antibody is bound to a surface with subsequent antigen binding followed by the addition of a second labeled antibody that reacts with the initial antigen/antibody complex.

Sothern blotting
A Southern blot is a method used in molecular biology for detection of a specific DNA sequence in DNA samples. Southern blotting combines transfer of electrophoresis-separated DNA fragments to a filter membrane and subsequent fragment detection by probe hybridization.

Western blotting
Membrane-based assay in which proteins are separated by electrophoresis, which is followed by transfer to a membrane and probing with a labeled antibody.

Immunochemical Problems

1. What is the difference between Southern and Northern blotting?

2. What are the components of an immunoglobulin G antibody molecule.

3. List three binding forces that act to produce antigen/antibody binding.

4. Explain how the following factors affect antigen/antibody binding
    a. Addition of a linear polymer
    b. Ion species
    c. Precipitin reaction

5. For each of the following qualitative immunochemical techniques, state the principle and clinical use:
    a. Immunofixation
    b. Western blotting

6. The energy of interaction of a single antibody-combining site and its corresponding epitope on the antigen is referred to as:
    a. sensitivity
    b. specificity
    c. immunogenicity
    d. Affinity
    Answer d

7. True or False: The addition of a linear polymer such as polyethylene glycol to a mixture of antigen and antibody causes decrease in the rate of immune complex formation and precipitation.
    Answer: False, the rate increases.

8. Which of the following substances does glucose-6-phosphate dehydrogenase bind to in the EMIT assay?
    a. Coenzyme
    b. Substrate
    c. Antigen
    d. Antibody
    Answer: c

9. Which of the following describes the composition of the two particles used in the luminescent oxygen channeling immunoassay (LOCI)?
    a. High-energy fluorescent radiation bead and a low-energy fluorescent radiation bead coated with luminal
    b. One synthetic bead coated with the enzyme alkaline phosphatase and a second synthetic bead coated with the substrate 4-nitrophenylphenol
    c. One paramagnetic particle coated with anti IgG antibody and a second paramagnetic particle w coated with an IgG receptor antigen
    d. One synthetic bead coated with Streptavidin that contains photosensitive dye and a second synthetic bead coated with a binding partner specific for the method and contains a chemiluminescent dye
    Answer: d

10. The antibody that binds to hapten-enzyme in the EMIT assay results in which of the following?
    a. Enhancement of enzyme activity
    b. Inhibition of enzyme activity
    c. Inhibition of the enzyme substrate complex to reach first order kinetics
    d. It has no effect on the reaction.
    Answer: b

11. Which of the following best describes a procedural difference between homogeneous and heterogeneous immunoassays?
    a. There is no physical separation of bound from free fractions in a homogeneous immunoassay whereas in a heterogeneous immunoassay you must separate bound forms from free forms.
    b. A homogeneous immunoassay usually takes longer to perform because you must physically separate the bound from free fractions whereas in a heterogeneous immunoassay the time to perform the assay is shorter because you do not have to separate bound forms from free forms.
    c. There is a physical separation of bound from free fractions in a homogeneous immunoassay.
    d. Separation of bound and free fractions is optional for either type of immunoassays.
    Answer: a

CHAPTER 7
Analytical and Clinical Evaluation of Methods

Clinical Significance

The purpose of a clinical laboratory test is to deliver data on the pathophysiological condition of an individual patient to assist with diagnosis, therapy, or risk assessment for a disease. Internal quality control, or quality control (QC), is a statistical sampling approach performed by a laboratory on a regular schedule to assess performance of its measuring systems.

Definitions and Key Terms

Analyte
The substance being analyzed in an analytical procedure.

Analytical measurement range
The analyte concentration range over which measurements are within the declared tolerances for imprecision and bias; also referred to as reportable range.

Analytical sensitivity
The ability of an analytical method to assess small variations in the concentration of analyte.

Analytical specificity
The ability of an assay procedure to determine specifically the concentration of the target analyte in the presence of potentially interfering substances or factors in the sample matrix.

Assay comparison
Comparison of measurements by two methods that is carried out objectively using statistical procedures and graphics displays.

Bias
In an analytical method, the difference between the average value and the true value that is expressed numerically and is inversely related to the trueness.

Calibration
In relation to analytical methods, a function that describes the relationship between instrument signal and concentration of analyte.

Coefficient of variation
Coefficient of variation is calculated as the standard deviation (SD) divided by the mean and multiplied by 100 to express in percent.

Commutability
The equivalence of the mathematical relationships between the results of different measurement procedures for a reference material and for representative samples from healthy and diseased individuals.

Correlation coefficient
Measure of association between two variables.

Deming regression
Least-squares regression analysis taking measurement errors in both variables into account.

Difference plot
A bias plot that shows the dispersion of observed differences between the measurements of two methods as a function of the average concentration of the measurements.

Error model
A model of the error structure.

External quality assessment
Also called external quality control or proficiency testing; an assessment process in which samples that simulate patient specimens are received from an external organization and results compared to those from

Frequency distribution
A distribution of the frequency (ordinate) as a function of the variable value (abscissa), that is, a histogram of absolute or relative frequencies.

Gaussian probability distribution
Bell shaped relative frequency distribution described under basic statistics.

Levey-Jennings chart
A graphical display with observed control values plotted on the y-axis and time, in days, shown on the x-axis. The y-axis shows the target value with plus and minus 1, 2, and 3 standard deviations.

Likelihood ratio
The probability of occurrence of a specific test value given that the disease is present divided by the probability of the same test value if the disease was absent.

Linearity
Range of values for which there is a linear relationship between concentration and signal.

Limit of detection
An assay characteristic defined as the lowest value that significantly exceeds the measurements of a blank sample.

Matrix
In relation to analytical methods, human serum that contains analytes.

Mean
Arithmetic average of variables.

Median
Equal to the 50th percentile of a set of variables.

Negative predictive value
The proportion of subjects with a negative test who do not have the disease.

Odds ratio
The probability of the presence of a specific disease divided by the probability of its absence.

Ordinary least-squares regression (OLR) analysis
A method used to estimate the unknown parameters in a linear regression assessment performed to minimize the sum of squared vertical distances between observed responses and responses predicted by linear approximation.

Population
In relation to analytical methods, the complete set of all observations that might occur as the result of performing a particular procedure according to specified conditions.

Positive predictive value
The proportion of subjects with a positive test who have the disease.

Precision
The closeness of agreement between independent results of measurements obtained under stipulated conditions. Usually expressed as the standard deviation.

Prevalence
The frequency of disease in the population examined.

Random error
Error that arises from imprecision of measurement of the type that is described by a Gaussian distribution.

Reference measurement procedure
A procedure of highest analytical quality that has been shown to yield value in assessing the veracity of other measurement procedures for the same quantity.

Regression analysis
A statistical analysis that compares measurement relations between two analytical methods.

Specificity
The proportion of subjects without disease who have a negative laboratory test result.

Sigma metric
A value that expresses the variation in performance of a procedure relative to the allowable variability in results to be suitable for medical decisions expressed in SD units. Six-sigma performance means that six SDs of measurement procedure variation fit within the allowable limits for acceptable performance.

Standard deviation (SD)
A statistical value that estimates the dispersion of replicate values around a mean value. A SD assumes a

Gaussian distribution of values that is typically observed for numeric quality control results. Standard deviation is calculated as the square root of the sum of squared deviations from the mean divided by number of variables minus one.

Student t distribution
Distribution related to the Gaussian distribution given a limited sample size.

Systematic error
Error in measurement that arises from calibration bias or no specificity of an assay and, in the course of a number of analyses of the same analyte, remains constant (y-intercept deviation from zero) or varies in a proportional way (slope deviation from unity) based on the analyte concentration.

Traceability
In relation to analytical methods, a concept based on a chain of comparisons of measurements that lead to a known reference value done to ensure reasonable agreement between measurements of routine methods.

Uncertainty
A parameter associated with the result of a measurement that characterizes the dispersion of the values that could reasonably be attributed to the measure; more briefly, uncertainty is a parameter characterizing the range of values within which the value of the quantity being measured is expected to lie.

Westgard Rules
The Westgard rules are a set of statistical patterns, each being unlikely to occur by random variability, thereby raising a suspicion of faulty accuracy or precision of the measurement system.

- 1_{2s}
    - One measurement exceeds two standard deviations either above or below the mean of the reference range.
    - Measures: Inaccuracy and/or imprecision

- 1_{3s}
    - One measurement exceeds three standard deviations either above or below the mean of the reference range.
    - Measures: Inaccuracy and/or imprecision

- 2_{2s}
    - Two consecutive measurements exceed two standard deviations of the reference range, and on the same side of the mean.
    - Measures: Inaccuracy.

- R_{4s}
    - Two measurements in the same run have a standard deviation difference (such as one exceeding 2 standard deviations above the mean, and another exceeding 2 standard deviations below the mean).
    - Measures: Imprecision.

- **4$_{1s}$**
  - Four consecutive measurements exceed 1 standard deviation on the same side of the mean.
  - Measures: Inaccuracy.
- **10$_x$**
  - Ten consecutive measurements are on the same side of the mean.
  - Measures: Inaccuracy.

Analytical and Clinical Evaluation of Methods Problems

1. A 1_{3s} quality control rule violation is defined as:
   a. One control result that exceeds ±3 SDs.
   b. One control result that exceeds ±1.3 SDs
   c. Three consecutive controls result that exceed ±3 SS.
   d. One control result that exceeds ±4SDs.
   Answer: a

2. A graph that shows the difference between your laboratory result and a comparative method results along the x-axis and the percent difference along the y-axis is referred to as which graph?
   a. Levy-Jennings graph
   b. Linear regression graph
   c. CUSUM graph
   d. Bias graph
   Answer: d

3. Which of the following is an example of insignificant type data?
   a. Numbers applied to numeric variables for example, cholesterol result of 150 mg/dL
   b. Numbers applied to nonnumeric variables for example blood types could be coded as group O is equal to 1, group A is equal to 2 etc.
   c. Numbers that are discrete and ordered for example, 4 = most severe injury, 2 = a moderately severe and 1 = a minor injury
   d. Arithmetic mean
   Answer: b

4. Which statement listed below is a reason to reject an analytical run?
   a. One control result exceeds above +1 standard deviation and the second control result exceeds -1 standard deviation from the mean.
   b. Two consecutive control results exceed 2 standard deviations above or below the mean.
   c. Four control results steadily increase in value but are less than ±1 standard deviation from the mean.
   d. Three consecutive control results exceed plus or minus 1 standard deviation from the mean.
   Answer: b

5. Which of the following practices is inappropriate when establishing quality controls ranges?
   a. Exclusion of any quality control results greater than ±2 standard deviation from the mean
   b. Compare your data to the manufacturer of the quality control material.
   c. Gather quality control data for a long period of time if necessary.
   d. Using control results from all shifts on which the assay is performed
   Answer: a

6. Which of the following statistics is used to qualify the strength of the relationship between two variables?
   a. z- value
   b. standard error of the difference
   c. Confidence interval
   d. Correlation coefficient
   Answer: d

Analytical and Clinical Evaluation of Methods

7. A significant limitation of least squares regression analysis is:
    a. A large confidence interval may result is false positives.
    b. Outliers will not impact the results.
    c. It may result in an incorrect decision of analytical reproducibility.
    d. Nonlinearity of the data will affect both the slope value and y-intercept.
    Answer: d

8. Linearity is useful for assessing:
    a. Slope and bias of an analytical method.
    b. A large correlation coefficient.
    c. Intercept and p value of analytical method.
    d. Slope and intercept of an analytical method.
    Answer: d

9. A glucose result that is greater than the lower limit of detection of the method should be reported as which of the following?
    a. Zero
    b. Less than the detection limit value
    c. The actual concentration value given by the analyzer or as calculated by the operator
    d. Greater than the detection limit value
    Answer: d

10. The results of a precision study urea nitrogen are shown below. What is the percent coefficient of variation for the study?
    Mean = 25 mg/dL
    Standard deviation = 1.5 mg/dL
    Variance = 2.0 mg/dL
    a. 1.5%
    b. 2.0%
    c. 10.0%
    d. 6.0%
    Answer: d

CHAPTER 8
Molecular Diagnostics

Clinical Significance

Molecular diagnostics and its parent field, molecular pathology, examine the origins of disease at the molecular level, primarily by studying nucleic acids. Deoxyribonucleic acid (DNA, which contains the blueprint for constructing a living organism, is the centerpiece for research and clinical analysis. Molecular pathology is an outgrowth of the enormous amount of successful research in the field of molecular biology that has discovered the basic biologic and chemical processes of how a living cell function (12-15).

Molecular biology is now used for clinical diagnosis and the development and use of therapeutics. The field of molecular diagnostics is an important and exciting area that is going to have an even greater impact on medicine in the future. As an increasing number of diseases are characterized at the molecular (e.g., nucleic acid and protein) level, new therapeutics and diagnostics specifically targeting these molecular changes will continue to emerge (1, 16, 17).

Definitions and Key Terms

Alleles
Different variants of a gene.

Annotation
Biologic information attached to a genomic sequence.

Assembly
Reconstruction of short-sequence reads on a scaffold of reference DNA. Autosome A non–sex chromosome.

Base pair
Complementary base pairing of guanine and cytosine and adenine and thymine through hydrogen bonding.

Centromere
The region of a chromosome between the small and long arms that is used during mitosis to attach the spindle fibers.

Chromatin
A structure in the eukaryotic nucleus composed of DNA and histone proteins.

Chromosome
A nucleic acid structure bound by protein containing all or a portion of an organism's genetic information.

Codon
A three-base code found in messenger RNA, used in translation to specify the amino acid to be incorporated in the growing polypeptide chain.

Copy number variant (CNV)
A structural variant of a large region of the genome that has been deleted or duplicated.

Deletion
A DNA sequence that is missing in one sample compared with another. Deletions may be as small as one nucleotide or as large as an entire chromosome.

Deoxyribonucleic acid (DNA)
Repository of the genetic material in an organism.

DNA methylation
Addition of a methyl group to a cytosine, used to regulate gene expression.

Epigenetics
Modifications that affect DNA packaging and accessibility without changing the DNA sequence.

Eukaryote
An organism that contains a nucleus and whose DNA is in chromosomes.

Exon
Portion of a gene that codes for protein.

Exonuclease
An enzyme that cuts in the middle of a stretch of DNA at a specific sequence of bases.

FASTQ file
A text output file of sequencing reads in a run, along with the quality scores of each position.

Fusion
A translocation, inversion, large deletion, or large duplication resulting in a hybrid gene formed from originally separate genes.

Gene
DNA structure that codes for the production of a protein.

Gene expression
Synthesis of a gene product in the form of an RNA or protein from a gene.

Genetic code
Collection of three base sequences called codons used to translate the nucleic acid information into amino acids during protein synthesis.

Genome
The complete genetic content of an organism.

Genotype
The unique genetic information of an organism or cell.

Heteroplasmic
A mixture of more than one type of mitochondrial sequence in one cell.

Histones
Basic proteins that bind to DNA to form nucleosomes, basic units of chromatin.

Indels
Commonly refers to as either an insertions or deletions or a combination of the two.

Insertion
An extra DNA sequence that is present in one sample compared with a reference sequence.

Intergenic
DNA sequence between genes.

Intron
DNA between exons that do not code for protein and are removed by RNA splicing.

Micro RNA (miRNA)
Small RNAs that are not translated and function to regulate translation.

Missense variant
A nucleotide substitution that changes a codon to the code for a different amino acid.

Mitochondrial DNA
Genetic information found uniquely in mitochondria that codes for proteins involved with energy production.

Mutation
A disease-causing sequence variation.

Nucleases
Enzymes that cleave DNA to smaller fragments.

Nucleic Acid
A polymer made of nucleotides consisting of a sugar, a phosphate group, and a nitrogenous base.

Nucleosome
A unit of DNA that contains 146 bases and is wound around 8 histone proteins.

Nucleotide
A basic unit of DNA and RNA consisting of a sugar, a phosphate group, and a nitrogenous base.

Phenotype
The physical and biologic characteristics of a genotype.

Phred score
Estimate of the error probability for a base called in DNA sequencing. It is represented as a Q score; the higher number represents the probability of a correct call.

Plasmid
An extrachromosomal ring of double-stranded closed DNA found in bacteria.

Polymerase
A protein that is able to connect individual molecules into a single strand.

Promoter
A DNA sequence usually upstream from the start of transcription of the gene that regulates its expression.

Pseudogene
A genetic element that does not code for a functional gene product, usually because of accumulated sequence variations.

Purine
A double-ring structure that constitutes the foundation for adenine and guanine, two of the bases in DNA.

Pyrimidine
A single ring structure that constitutes the foundation for thymine and cytosine, two of the bases in DNA and uracil and one of the bases of RNA.

Recombination
The rearrangement of DNA through the breaking and rejoining of DNA strands to create a different sequence.

Ribonucleic acid (RNA)
A polymer made of ribonucleotides consisting of ribose, a phosphate group, and four bases—guanine, adenine, cytosine, and uracil.

Ribosome
A large structure that functions to convert messenger RNA to protein.

Single nucleotide polymorphism (SNP)
A single nucleotide variant that occurs in a population at a frequency of at least 1%.

Single nucleotide variation (SNV)
A single nucleotide variant that may be benign or cause disease.

Spliceosome
A large structure that functions to remove introns from a heterogeneous RNA and splice the remaining exons together to form a messenger RNA.

Structural variation
A region of DNA greater than 1000 bases in size that is inverted, translocated, inserted, or deleted.

Synonymous variant
A nucleotide change that results in no change to the predicted amino acid sequence.

Telomere
A repetitive DNA sequence at the end of a chromosome.

Transcription
The biologic process that copies the information in DNA into RNA.

Translation
The biologic process that uses ribosomes to convert RNA to protein.

Variant
A disease-causing sequence variation. Historically the term has been interchangeable with mutation to describe any change in a DNA sequence regardless of its relation to disease causation.

Molecular Principles Problems

1. Compare and contrast the structure and function of DNA and RNA.

2. Identify five different types of RNA in the cell and describe their functions.

3. Describe the central dogma of molecular biology.

4. Which of the following is TRUE about gel electrophoresis of DNA?
    a. DNA of smaller molecular weight will travel far from the point of application.
    b. DNA migrates toward the cathode.
    c. Increasing the voltage increases the resolution of DNA fragments separated in the gel.
    d. DNA electrophoresis requires polyacrylamide as a matrix.
    Answer: c

5. Restriction endonucleases function by:
    a. Linking 2 pieces of DNA from different sources.
    b. Adding a phosphate group onto one end of the DNA.
    c. Cutting the DNA at specific nucleotide sequences.
    d. Breaking the DNA randomly to create fragments of manageable size.
    Answer: c

6. All of the following are clinical applications of FISH EXCEPT:
    a. Detection of chromosome microdeletions.
    b. Detection of oncogenes in tumor cells.
    c. Monitoring patients with sex-mismatched bone marrow transplants for engraftment.
    d. Detecting DNA from tumor-causing viruses in cytologic specimens.
    Answer: b

7. RNA differs from DNA in what way?
    a. RNA Contains thymine instead of uracil.
    b. RNA Contains an H group instead of an OH group at its number 2 carbon.
    c. RNA Is normally single-stranded.
    d. RNA Is used to carry the genetic information of most viruses.
    Answer: b

8. The protein product derived from the DNA sequence below contains which of the following amino acid sequences?

9. DNA: 3' TACTTTCGCGGAACT 5'
    a. Tyrosine-phenylalanine-arginine-glycine-threonine
    b. Tyrosine-phenylalanine-arginine-glycine
    c. Methionine-lysine-alanine-proline
    d. Methionine-lysine-alanine-proline-tryptophan
    Answer: c

CHAPTER 9
Molecular Techniques

Clinical Significance

Molecular methods use techniques for diagnosis in the acute phase of a disease, before the appearance of antibodies exactly like antigen detection. The expansion and power of molecular diagnostics is empowered by techniques that modify, amplify, detect, differentiate, and sequence nucleic acids. Molecular diagnostic techniques in clinical laboratory science is getting faster, better, and less expensive (16-22).

Definitions and Key Terms

Adapter
Oligonucleotides that are ligated to library fragments in order to provide consensus priming sites.

Asymmetric PCR
A version of PCR that preferentially amplifies one strand of the target DNA.

Comparative genomic hybridization
A technique in which reference and test DNA are respectively labeled with green and red fluorochromes. Abnormalities are detected by changes in the green-to-red ratio.

Dideoxy-termination sequencing
A method of DNA sequencing based on the selective incorporation of chain-terminating dideoxynucleotides by DNA polymerase during in vitro DNA replication.

Digital polymerase chain reaction
A modification of PCR with the sample separated into many partitions so that some partitions have no template and others have more.

DNA microarray
An array of microscopic spots of different DNA molecules attached to a solid surface. Also known as DNA chips or biochips.

Flow cytometry
A technique for counting cells or particles suspended in fluid as they flow one at a time past a focus of exciting light.

Fluorescence in situ hybridization (FISH)
A genetic mapping technique using fluorescent tags for analysis of chromosomes and genetic abnormalities.

High-resolution melting analysis (HRMA)
A simple method for genotyping that associates fluorescence against temperature using a dye that detects double-stranded but not single-stranded DNA.

Loop-mediated isothermal amplification (LAMP)
A single tube technique for the amplification of DNA that uses a single temperature incubation.

Massively parallel sequencing (MPS)
Sequencing of many fragments of DNA simultaneously.

Melting temperature (Tm)
The temperature at which half of the DNA strands are in the duplex form.

Multiplex ligation-dependent probe amplification (MLPA)
A variation of the multiplex PCR that assesses the copy number of several targets by permitting multiple targets to be amplified with only a single primer pair.

Oligonucleotide ligation assay (OLA)
A technique for determining the presence or absence of a specific nucleotide pair within a target gene, often indicating whether the gene is wild type (normal) or mutant (defective).

Polymerase chain reaction (PCR)
An in-vitro method for exponentially amplifying DNA.

Pyrosequencing
A method of DNA sequencing based on the "sequencing by synthesis" principle that relies on the detection of pyrophosphate release on nucleotide incorporation.

Real-time PCR
Observation of PCR during amplification at least once each cycle.

Rolling circle amplification (RCA)
A probe amplification method with a linear probe that is ligated to form a circle in the presence of template. The circle is replicated continuously by a polymerase and one or more primers.

Strand displacement amplification (SDA)
An amplification technique that uses two types of primers, DNA polymerase, and a restriction endonuclease to exponentially produce single-stranded amplicons asynchronously.

Transcription-mediated amplification (TMA)
An amplification method that uses RNA polymerase and DNA reverse transcriptase to produce RNA amplicon from a target nucleic acid. TMA is used to amplify both RNA and DNA.

Whole-genome amplification (WGA)
A nonspecific amplification technique that produces an amplified product representative of the initial starting material (whole genome).

Molecular Techniques Problems

1. State the difference between target amplification and signal amplification.

2. Describe the different ways that fluorescence can be generated during real-time PCR.

3. Describe the differences between transcription mediated amplification (TMA) and the PCR.

4. Describe how digital PCR is different from regular PCR.

5. Describe one method for direct quantification of nucleic acids without amplification.

6. What is high-resolution melting analysis?

7. Describe the kind of controls needed for the PCR.

8. Which of the following is NOT a component of a nucleotide?
    a) 5' carbon sugar
    b) Amino acid
    c) Nitrogenous base
    d) Phosphate group
    Answer: c

9. Suppose that a laboratory wanted to separate small DNA fragments that were 10 base pairs (bps) apart in size. The optimal technique to use for the separation is:
    a) Agarose gel electrophoresis.
    b) Polyacrylamide gel electrophoresis.
    c) Pulse-field gel electrophoresis.
    d) Denaturing gel electrophoresis.
    Answer: c

10. Which of the following fragments would be generated when the following sequence is cut by SmaI?
    5' TACCCCGGGGGCAATTCCCGGGAGATTCCCGGGAACTC 3'
    a) One 3 bp fragment, two 11 bp fragments, and one 13 bp fragment
    b) One 4 bp fragment, one 10 bp fragment, one 11 bp fragment, and one 13 bp fragment
    c) Two 19 bp fragments
    d) One 6 bp fragment, one 8 bp fragment, one 11 bp fragment, and one 13 bp fragment
    Answer: b

11. Real-time PCR differs from traditional PCR in that:
    a) Primers are not necessary.
    b) The amplicon is detected as the reaction progresses.
    c) The amplicon is detected on a special fluorescent gel.
    d) The reaction requires hybridization probes.
    Answer: d

12. More nonspecific binding of a probe to DNA sequences in a sample is encouraged when:
    a) The size of the probe is increased.
    b) The salt concentration is decreased.
    c) The temperature is decreased.
    d) The temperature is increased.
    Answer: c

13. Labeled antibodies are used as probes to detect proteins in:
    a) Hybrid capture assay.
    b) Northern blot.
    c) Southern blot.
    d) Western blot.
    Answer: d

14. A suitable method to use for quantitation of HIV in patient blood is:
    a) bDNA.
    b) DNA sequencing.
    c) SDA.
    d) Southern blot.
    Answer: a

CHAPTER 10
Therapeutic Drug Monitoring (TDM)

Key Equations
- Half-life $(t_{1/2}) = \ln 2 \div k_{el}$
- $Abs = 2 - \log|\%T|$
- Dose (mg/kg) = Plasma Concentration x Volume of Distribution
- Volume of Distribution (L/kg) = Dose (mg/kg) ÷ Plasma Concentration (mg/L)

Clinical Significance

Knowledge about and understanding of the pharmacokinetics and pharmacodynamics of each drug allow the CLS and the clinician an opportunity to provide the patient with the best drug dosage personalized care. Therapeutic drug monitoring involves the coordinated effort to accurately monitor circulating drug concentrations in serum, plasma, and whole blood specimens (23, 24).

The main purposes for TDM are to:
1. Ensure drug dosage is within a range that produces maximal therapeutic benefit known as the therapeutic range.
2. Identify when the drug is outside the therapeutic range, which may lead to drug inefficacy or toxicity.

For most drug therapies, safe and effectual dosage regimens for the majority of the population have been established, and TDM is unnecessary. However, for some therapeutic drugs, there is a narrow window between therapeutic efficacy and toxicity. Therefore, careful monitoring and appropriate dosage adjustments are necessary to maintain therapeutic concentrations.

Definition and Key Terms

Antiepileptic
A substance to prevent or alleviate seizures.

Applied pharmacokinetics
The use of pharmacokinetic principles to guide the clinical application of therapeutic drugs.

Beta-blocker
A drug that induces adrenergic blockade at either Pl-or P2-adrenergic receptors or at both.

Bioavailability
The fraction of a drug absorbed into the systemic circulation relative to the same drug administered intravenously.

Biotransformation
The chemical alteration of a xenobiotic within the body that generally enhances the xenobiotic's aqueous solubility and excretion.

Dose-response relationship
The relationship between the dose of an administered drug and the response of the organism to the drug.

Drug half-life
The time required for one-half of an administered drug to be lost through metabolism and elimination.

Drug interactions
The effects of one drug on the intestinal absorption, metabolism, or action of another drug.

Enzyme induction
Increased synthesis of an enzyme in response to an inducer or other stimulus.

First-pass effect
Extensive metabolism of a drug with a high hepatic extraction rate by the liver before it reaches the systemic circulation.

Generic drug
A drug not protected by a trademark. Also, the scientific name as opposed to the proprietary, brand name.

Immunophilin
A generic term for an intracellular protein that binds immunosuppressive drugs such as cyclosporine, FK 506, or rapamycin.

Immunosuppressant
An agent capable of suppressing immune responses.

Pharmacodynamics
The study of the biochemical and physiological effects of drugs and the mechanisms of their actions, including the correlation of actions and effects of drugs with their chemical structure; also, such effects on the actions of a particular drug or drugs.

Pharmacokinetics
The activity or fate of drugs in the body over a period of time, including the processes of absorption, distribution, localization in tissues, biotransformation, and excretion.

Therapeutic drug monitoring
The process of using drug concentration or other pharmacodynamics biomarkers to guide drug dosing.

Toxicology
Subdiscipline of pharmacology concerned with adverse effects of chemicals on living systems.

Volume of distribution (Vd)
Volume of distribution is a theoretic concept that relates the amount of drug in the body (dose) to the concentration (C) of drug that is measured plasma. Volume of distribution is the volume of fluid

"apparently" required to contain the total-body amount of drug homogeneously at a concentration equal to that in plasma.

Xenobiotics
A chemical substance foreign to the biological system.

TDM Problems

1. What are the characteristics of a drug that make therapeutic drug monitoring essential?

2. List the factors that influence the absorption of an orally administered drug.

3. The fractional extent to which a dose of drug reaches its site of action is referred to as:
    a. Elimination half-life.
    b. Phase I reaction.
    c. First- pass effect
    d. Bioavailability.
       Answer: d

4. Which of the following best reflect the four correct pharmacological parameters that determine serum drug concentration?
    a. Absorption, anabolism, bioactivation, excretion
    b. Equilibration, biotransformation, reabsorption, elimination
    c. Absorption, distribution, metabolism, excretion
    d. Ingestion, conjugation, metabolism, elimination
       Answer: c

5. Which of the following is an example of an antiarrhythmic drug that has a metabolite with very similar same pharmacological action?
    a. Procainamide
    b. Phenobarbital
    c. Quinidine
    d. Caffeine
       Answer: a

6. Which of the following compounds selectively inhibits proliferation and activation of $CD4^+$ T cells?
    a. Digoxin
    b. Beta hCG
    c. Progesterone
    d. Cyclosporine A
       Answer: d

7. Which of the following is the correct definition of first-pass effect of a drug?
    a. When a drug is excreted into the urine
    b. When a drug is metabolized in the gut wall and liver before it reaches the systemic circulation
    c. The time it takes for the plasma concentration of drug in the body to be reduced by 50%
    d. The point when the total amount of drug in a human does not change over multiple loading does intervals
       Answer: b

8. The goal of therapeutic drug monitoring (TDM) is to:
    a. Provide the patient with the least amount of drug to effect the best outcome.
    b. Provide the patient with the cheapest formulation of the drug to effect the best outcome.
    c. Provide the patient with the best estimate of a dosage of drug to effect the best outcome.
    d. Provide the patient with the optimum dosage of drug to effect the best outcome.
    Answer: d

9. The pharmacological response of a drug is initiated when the drug is in which form?
    a. Bound to fatty acids
    b. Bound to proteins
    c. Bound to carbohydrates
    d. Free
    Answer: d

10. Plasma clearance of a drug is defined as:
    a. The volume of plasma from which all drug appears to be removed in a given volume of blood (e.g., mL/mL).
    b. The volume of plasma from which all drug appears to be removed in a given time (e.g., mL/min).
    c. The amount of drug removed from urine in a given time (e.g., mg/min).
    d. The amount of time it takes for a drug to reach seven half-lives.
    Answer: b

11. Steady state of a drug is typically achieved after:
    a. Approximately seven half-lives.
    b. Approximately two half-lives.
    c. The drug is completely absorbed from the gut.
    d. The drug first begins to appear in the urine.
    Answer: a

12. Which of the following drugs is used as an immunosuppressant in liver transplants?
    a. Thiocyanate
    b. Tacrolimus
    c. Cyclosporine
    d. Primidone
    Answer: b

13. A drug that is administered into the subarachnoid space or ventricles (intrathecal) is referred to as:
    a. Oral.
    b. Topical.
    c. Parenteral.
    d. Enteral.
    Answer: c

Therapeutic Drug Monitoring (TDM)

14. Which of the following compounds is an anticonvulsant used to control tonic-clonic (grand mal) seizures?
    a. Digoxin
    b. Acetaminophen
    c. Lithium
    d. Phenytoin
    Answer: d

15. Which of the following compounds relaxes the smooth muscles of the bronchial passages?
    a. Acetaminophen
    b. Lithium
    c. Phenytoin
    d. Theophylline
    Answer: d

16. Which of the following compounds is a cardiac glycoside that is used in the treatment of congenital heart failure and arrhythmias by increasing the force and velocity of myocardial contraction?
    a. Digoxin
    b. Carbamazepine
    c. Amikacin
    d. Phenobarbital
    Answer: a

17. Carbamazepine, primidone, and valproic acid are all examples of:
    a. Antiarrhythmic drugs.
    b. Antidepressant drug.
    c. Antibiotics.
    d. Antiepileptic drugs.
    Answer: d

18. Depakene or Depakote are proprietary names for which of the following compounds?
    a. Phenobarbital
    b. Dilantin
    c. Valproic acid
    d. Carbamazepine
    Answer: c

19. Following the entrance of a drug into the vascular compartment, it will be carried to all organs and fluid compartments. This statement describes which of the following?
    a. Digestion
    b. Distribution
    c. Metabolism
    d. Reperfusion
    Answer: b

20. Phase I biotransformation includes:
    a. Attachment of a hydroxyl group.
    b. Conjugation with glucuronic acid.
    c. Conjugation with sulfate.
    d. Conjugation with glutathione.
    Answer: a

21. Which of the following compounds is the major metabolite of primidone?
    a. Digoxin
    b. Theophylline
    c. N-acetylprocainamide
    d. Phenobarbital
    Answer: d

22. A blood sample for a trough drug level should be drawn:
    a. During the distribution phase of the drug.
    b. 1.5 hours after drug administration.
    c. Shortly before drug administration.
    d. During phase I biotransformation.
    Answer: c

23. Which of the following are the toxic effects of phenobarbital?
    a. Atrial fibrillation, hyperactive, and very fast breathing rates
    b. Severe tiredness/dizziness, inability to wake up, and very slow breathing rates
    c. Ototoxicity and possible nephrotoxicity
    d. Spontaneous hemorrhage or life-threatening infections
    Answer: b

24. Clearance (CL) of a drug is defined as which of the following?
    a. The rate of elimination by all routes normalized to the concentration of drug, in some biological fluids
    b. A measure of the apparent space the body has available to contain the drug
    c. The time required for an amount of drug in blood to decline to one-half its measured value
    d. Drug dose reaching the systemic circulation following administration by any route
    Answer: a

25. Which of the following is bactericidal for *Streptococci, Corynebacterium, Clostridia, Listeria,* and *Bacillus* species?
    a. Tobramycin
    b. Cyclosporine
    c. Vancomycin
    d. Methotrexate
    Answer: c

Therapeutic Drug Monitoring (TDM)

26. Phenobarbital is classified as a (an):
    a. Cardiac glycoside.
    b. Anti-arrhythmic.
    c. Analgesic, antipyretic, and anti-inflammatory.
    d. Sedative and hypnotic.
    Answer: d

27. Calculate the half-life ($t_{1/2}$) in hours using the equation below.
    y = -0.09526x + 0.5848 R^2 = 0.968   Answer: 7.27 hrs

28. Using a spectrophotometer, phenobarbital the % Transmittance was determined. Using the data below and determine the concentration of phenobarbital for a patient who's %T is 16.6%

Concentration (mg/dL)	% T
0	100
0.1	40
0.2	15
0.3	7
0.4	4
0.5	2

    Answer: 0.21 mg/dL

CHAPTER 11
Amino Acid and Proteins

Clinical Significance

Amino acids are molecules that combine to form proteins thus amino acids are the building blocks of life. The chemical properties of the amino acids of proteins determine the biologic activity of the protein. Growth, repair, and maintenance of all cells are dependent on amino acids. The chemical properties of the amino acids of proteins determine the biologic activity of the protein. There are 20 amino acids used by the human body. Selenocysteine is recognized as the 21st amino acid and Pyrrolysine is the 22nd. Essential amino acids cannot be made by the body. As a result, they must come from food. The nine essential amino acids are: histidine, isoleucine, leucine, lysine, methionine, phenylalanine, threonine, tryptophan, and valine. Nonessential amino acid can be produced *in-vivo*. Nonessential amino acids include: alanine, arginine, asparagine, aspartic acid, cysteine, glutamic acid, glutamine, glycine, proline, serine, and tyrosine.

Proteins play key roles in every cell throughout the human body. Some of their major functions include catalyzing biochemical reactions as enzymes, transporting metals such as iron and copper, acting as receptors for hormones providing structure and support to cells, and participating in immune response as antibodies. Proteins serve a key role in transport, synthesis, storage, and clearance of substances and are classified by their function:

- **Immune defense**
    - Immunoglobulins used for the elimination of foreign antigens
        - IgA, IgG, IgM, IgD, and IgE

- **Acute phase reactants (APR)**
    - Protein associated with inflammation
        - Negative APR: Albumin, prealbumin, and transferrin
        - Positive APR: Alpha-1 globulin, Alpha-2 globulin, Beta globulin, Delta globulins
- **Transport**
    - Proteins used to bind and transport substrates and substances
        - Prealbumin, albumin, ceruloplasmin, Alpha-2-macroglobulin, transferrin
- **Coagulation**
    - Proteins aiding in clot formation and are a part of the complement cascade
        - D-dimer, Fibrinogen, Complement C3 and C4

Aminoacidopathies

Aminoacidopathies are a class of inherited errors of metabolism in which there is an enzyme defect that inhibits the body's ability to metabolize certain amino acids. The abnormalities exist either in the activity of a specific enzyme in the metabolic pathway or in the membrane transport system for amino acids. More than one-hundred diseases have been identified that result from inherited errors of amino acid metabolism. Aminoacidopathies cause medical complications due to the buildup of toxic amino acids or by-products of amino acid metabolism in the blood. Due to the severity of these complications, many states require newborn screening tests to aid in early diagnosis and initiation of therapies (25).

Types of Aminoacidopathies

- **Phenylketonuria**
    - Autosomal recessive genetic defect in the enzyme phenylalanine hydroxylase (PAH)
- **Tyrosinemia**
    - Disorder of tyrosine catabolism
        - Type 1 Caused by low level of fumarylacetoacetate hydroxylase
        - Type 2 Caused by deficiency of the enzyme tyrosine aminotransferase
        - Type 3 Caused by deficiency of 4-hydroxyphenylpyruvate dioxygenase
- **Alkaptonuria**
    - Inborn metabolic disease autosomal recessive
    - HGD gene-causes lack of enzyme homogentisate oxidase
- **Maple Syrup Urine Disease**
    - Results from absence or greatly reduced activity of alpha-keto-acid decarboxylase.
    - Blocks normal metabolism of leucine, isoleucine and valine
- **Isovaleric Acidemia**
    - Deficiency of isovaleryl CoA dehydrogenase
    - Prevents normal metabolism of leucine
- **Homocystinuria**
    - Inherited autosomal recessive disorder, lacks enzyme cystathionine Beta synthase for metabolism of methionine.
- **Citrullinemia**
    - Autosomal recessive disorder
        - Type 1 Lacks enzyme argininosuccinic acid synthase
        - Type 2 A mutation in gene for citrin protein
- **Argininosuccinic Aciduria**
    - Lack enzyme argininosuccinic acid lyase
- **Cystinuria**
    - Autosomal recessive defect in amino acid transport system
    - Inadequate reabsorption of cysteine in kidneys

Amino Acid Method of Analysis

Urinary amino acid analysis can be performed for screening purposes. To quantitate the amino acid, a 24-hour urine specimen should be preserved with thymol, a phenolic compound, or organic solvents for analysis. When identifying a particular category of amino acids, such as branched-alkyl chain amino acids or a single carboxylic acid, Thin Layer Chromatography (TLC) is sufficient. For a more general screening, column chromatography is essential. Amino acids can also be separated and quantitated by ion exchange chromatography, an HPLC reversed-phase system equipped with fluorescence detection, or capillary electrophoresis. Hyphenated instruments like LC-MS/MS is a highly specific and sensitive method for the measurement of amino acids (4, 6).

Plasma Proteins
Plasma proteins are the most commonly analyzed proteins in the clinical laboratory and can be divided into two major groups: albumin and globulins.

Types of Plasma Proteins
- Pre-Albumin
    - Named because it migrates before albumin in classic serum protein electrophoresis.
    - Transport protein for thyroxine and triiodothyronine can bind with retinol binding protein.
- Albumin
    - Albumin is synthesized in liver. It's a protein with highest concentration in plasma, responsible for 80% of osmotic pressure, buffers pH, binds various substances, transports fat soluble hormones, decreased levels of albumin can indicate disease.
    - Increase in blood concentration is associated with dehydration.
    - Decreased blood concentrations of albumin are most commonly associated with an acute inflammatory response as albumin is a negative acute-phase reactant. Liver and kidney disease may also result in low blood albumin concentrations.
    - High serum total protein but low albumin is usually seen in multiple myeloma.
- Globulin
    - Globulins consists of alpha1/alpha2/beta and gamma fractions, have many different functions
        - Alpha-1 globulin: 0.1 – 0.3 g/dl
        - Alpha-2 globulin: 0.6 – 1.0 g/dl
        - Beta globulin: 0.7 – 1.1 g/dl

Protein of Clinical Significance

- **Myoglobin**
    - The primary oxygen carrying protein found in striated skeletal and cardiac muscle. Myoglobin can reversibly bind to oxygen. Increase Concentration suggest muscle damage.
- **Cardiac Troponin**
    - The cells of the myocardium contain three types of troponin: troponin I, T, and C. Troponin C is expressed in skeletal muscle, while troponins I and T are expressed in cardiac myocytes and are clinically significant for diagnosis of myocardial infarction.
- **Brain Natriuretic Peptide (BNP)**
    - Natriuretic peptide family hormones that affect body fluid homeostasis and blood pressure
    - BNP is used as a market for congestive heart failure
- **Fibronectin**
    - Fibronectin is a glycoprotein used in cell adhesion, tissue differentiation, growth and wound healing. Fibronectin can be used as a nutritional marker.
- **Adiponectin**
    - Adiponectin is a fat hormone low levels correlate with increased risk of heart disease, type 2 diabetes and obesity.
- **Beta Trace Protein**
    - Beta Trace Protein is a biomarker of cerebrospinal fluid (CSF) leakage and impaired renal function.
- **Cystatin C**
    - Cystatin C is a cysteine proteinase inhibitor produced and destroyed at a constant rate and is used to assess glomerular filtration rate (GFR)
- **Amyloid**
    - Amyloid is an insoluble protein aggregate that is formed due to alterations in beta sheets. It is used as a biomarker for Alzheimer's Disease diagnosis.

Total Protein Analysis

There are four major types of globulins with individual properties and functions. Analysis of blood specimens will typically include four protein measurements: total protein, albumin, globulins, and the albumin-to-globulin (A/G) ratio. Albumin makes up 60% of the blood's protein while globulin makes up around 40%. The normal A/G ratio is 0.8 to 2.0. If, however, either the albumin is lowered, or globulins increase for various reasons, this ratio can be altered. This would be seen in chronic inflammatory diseases, multiple myeloma, and autoimmune diseases such as rheumatoid arthritis. For diagnosis, total protein may be measured if you have symptoms that could be caused by a liver or kidney problem. Total protein test and the A/G ratio may be associated with a liver panel or comprehensive metabolic panel that is done during a routine doctor visits for patients with an elevated risk of kidney and/or liver disease (26) (27).

Total Protein	6.1 – 8 g/dL
Albumin	3.2 – 5 g/dL
Globulin	2.2 – 4 g/dL

Methods of analysis for total protein

Total Protein abnormalities are used to determine nutritional status, kidney disease, liver disease, and many other conditions. Hyperproteinemia is an increase in total plasma proteins and is the result of dehydration or types of overproduction (ex. myeloma.) Hypoproteinemia is a decrease in total protein which can be caused by excessive loss or decreased intake (like malnutrition.) There are several ways of analyzing the amount of protein and they are listed in the next chart.

Method of Analysis for Total Protein

Kjedahl
Kjeldahl's method measures the nitrogen content of proteins as ammonium ion by back titration following oxidation of proteins by sulfuric acid and heat. It assumes that proteins average 16% nitrogen by weight. Protein in grams per deciliter is calculated by multiplying protein nitrogen by 6.25. The Kjeldahl method is a reference method for total.

Biuret Assay
Biuret is a compound called carbamoylurea that chelates copper (II) ions. Cupric ions are used to complex with groups involved in peptide bond Violet chelate is formed and measured at λ540 nm

Dye Binding
Depends on ability of most proteins to bind to dyes. Bromcresol green (BCG) and Bromcresol purple (BCP) are anionic dyes that undergo a spectral shift when they bind albumin at acidic pH. BCP is more specific for albumin than BCG. Reaction of both dyes with globulins requires a longer incubation time than with albumin, and reaction times are kept at 30 seconds or less to increase specificity.

Folin–Lowry Method
The Folin–Lowry method uses both biuret reagent and phosphotungstic and molybdic acids to oxidize aromatic side groups of proteins. These, in turn, reduce the Copper (II) ion in the biuret reagent, increasing sensitivity.

Electrophoresis
Electrophoresis is the migration of charged molecules in an electric field. Increasing the strength of the electrical field by increasing voltage increases migration.

Capillary Electrophoresis
Capillary electrophoresis is a rapid automated procedure for separating serum or body fluid proteins. Proteins migrate based upon their charge/mass ratio inside a silica capillary tube.

Definitions and Key Terms

Acute-phase response
Body's response to injury or inflammation.

Amino acid
An organic compound containing both amino ($-NH_2$) and carboxyl (-COOH) functional groups.

Aminoaciduria
An excess of amino acids in the urine.

Bence-Jones proteins
Small light chains of immunoglobulin found in the urine.

Complement system
Complex system of proteins found in blood that combines with antibodies to destroy pathogenic bacteria and other foreign cells.

Essential amino acids
Amino acids that are not synthesized by humans and therefore are essential dietary constituents for maintaining health or growth.

Immunoglobulins
A family of proteins also known as antibodies that contain highly specific antigen-binding sites consisting of two identical heavy (H) chains encoded on chromosome 14 and two identical light (L) chains encoded on chromosome 2.

Isoelectric focusing
An equilibrium technique that is used to separate charge variants of proteins and is applied to the analysis of certain genetic variants of proteins.

Kwashiorkor
A form of protein-energy malnutrition produced by severe protein deficiency.

Marasmus
A form of protein-energy malnutrition predominantly due to prolonged severe caloric deficit.

Multiple myeloma
A cancer in which antibody-producing plasma cells grow in an uncontrolled and malignant manner.

Paraprotein
A monoclonal immunoglobulin produced in excessive amounts in disorders such as multiple myeloma.

Peptide
A compound consisting of two or more amino acids linked in a chain via peptide bonds.

Peptide bond
The amide bond formed between the carboxyl group of one amino acid and the amino group of another.

Protein
A polymer of amino acids linked by peptide bonds with a specific sequence that folds into a defined structure; any of a group of complex organic compounds that contain carbon, hydrogen, oxygen, nitrogen, and usually sulfur (the characteristic element being nitrogen).

Proteome
The total complement of proteins expressed by the genetic material of an organism under a given set of environmental conditions.

Amino Acid Problems

1. Name two amino acids that are not found in humans.

2. What is the function and clinical significance of the following proteins:
    a. Prealbumin
    b. Albumin
    c. Myoglobin
    d. Cardiac troponin
    e. Brain Natriuretic Peptide

3. Hyperalbuminemia is caused by what?
    a. Dehydration syndromes
    b. Liver disease
    c. Kidney stones
    Answer: a

4. List the reasons for determining a patient's albumin-to-globulin (A/G) ratio?

5. How is the biuret method based?
    a. The reaction of phenolic groups with Cu_2SO_4
    b. Coordinate bonds between Cu^{+2} and carbonyl and imine groups of peptide bonds.
    c. The protein error of indicator effect producing color when dyes bind protein
    d. The reaction of phosphomolybdic acid with protein
    Answer: b

6. Which protein is used to assess nutritional status?

7. Bromcresol Green (BCG) may be used in the dye-binding assay of:
    a. Lipoprotein
    b. Cholesterol
    c. 17-ketosteroids
    d. Albumin
    e. Estrogen
    Answer: d

8. High serum total protein but low albumin is usually seen in:
    a. Multiple myeloma
    b. Hepatic cirrhosis
    c. Glomerulonephritis
    d. Nephrotic syndrome
    Answer: a

9. Matching
    - Aminoacidopathy _____
    - Phenylketonuria _____
    - Tyrosinemia _____
    - Alkaptonuria _____
    - Maple Syrup Urine Disease _____

    a. A mutation in the HGD gene leads to a deficiency of the enzyme homogentisate oxidase (HGD)
    b. Cause severe medical complications, such as brain damage, due to the accumulation of toxic amino acids or their by-products in the blood and tissues.
    c. Results from an absence or greatly reduced activity of a complex of enzymes known as branched-chain α-ketoacid decarboxylase (BCKD).
    d. Diagnostic criteria include an elevated plasma tyrosine concentration and an elevated concentration of succinylacetone.
    e. The absence of activity of the enzyme phenylalanine hydroxylase

Answer
    - Aminoacidopathy: b
    - Phenylketonuria: e
    - Tyrosinemia: d
    - Alkaptonuria: a
    - Maple Syrup Urine Disease: c

CHAPTER 12
Non-Protein Nitrogen Compounds

Clinical Significance

Creatinine, urea, and uric acid are nonprotein nitrogenous metabolites that are cleared from the body via the kidney after glomerular filtration. Plasma or serum concentrations of NPN's are frequently used as indicators of kidney function and other conditions. In particular, creatinine is commonly used to estimate the glomerular filtration rate (GFR). Small molecular weight proteins such as cystatin C are also increasingly being used to estimate the GFR. The measurement of total protein and/or albumin in urine can give valuable information regarding the integrity of the glomerular filter and renal tubular function (27-34).

Nonprotein Nitrogen Compounds (NPNs)

NPNs are waste products formed in the body as a result of the degradation and metabolism of nucleic acids, amino acids, and proteins. Excretion of these compounds is an important function of the kidneys. The six non-protein nitrogen compounds are: amino acids, ammonia, urea, creatine, creatinine, and uric acid.

Urea

Urea is referred to as Blood Urea Nitrogen (BUN) and is produced when amino acids are metabolized as nitrogen is released and converted to urea and excreted as a waste product. BUN occur due to kidney damage, low fluid intake, exercise, high protein intake, certain drugs, heart failure, or intestinal bleeding. Low levels are seen when the person has a low nitrogen diet, liver damage, malabsorption, or poor diet.

Reference Range Values:
- Blood Urea Nitrogen (BUN): 3 – 20 mg/dL
- Urea (urine) 12 – 20 g/dL

The clinical applications are:
- Evaluate renal function
- Assess hydration status
- To assist in diagnosis of renal disease

The analytical methods used to elucidate the concentration of urea are:
- Enzymatic approaches are used frequently
- Urease coupled with glutamate hydroxylase

Elevated concentration of urea in blood is called azotemia. Elevated concentration of BUN with renal failure is uremia. Azotemia is referred to as pre-renal, renal or post renal azotemia. The BUN to creatinine ratio can aid in the diagnosis of disease.

Pre-Renal Azotemia:
BUN:Creatinine > 20:1
Pre-renal azotemia is a result of a reduction of renal blood flow. When less blood is delivered to the kidney; less urea is filtered. Contributing factors include congestive heart failure, shock, hemorrhage, dehydration. Prerenal conditions tend to elevate plasma urea, whereas plasma creatinine remains normal, causing a BUN:Creatinine ratio.

Renal Azotemia
BUN:Creatinine < 10:1
Renal azotemia is a result of a reduction in renal function which causes an increase in plasma urea concentration resulting in compromised urea excretion. Renal causes of elevated urea include acute and chronic renal failure, glomerular nephritis, tubular necrosis, and other intrinsic renal disease thus a low BUN:Creatinine ratio is observed.

Post-Renal Azotemia
BUN:Creatinine = 10:1 - 20:1
Post-renal azotemia is be due to an obstruction of urine flow in the urinary tract by renal calculi, tumors of the bladder or prostate, kidney stones, or severe infection. A BUN:Creatinine ratio with an elevation of both BUN and creatinine is usually seen in postrenal conditions.

Creatinine

Creatinine is formed from creatine and creatine phosphate in muscle and is excreted into the plasma at a constant rate related to muscle mass. Plasma creatinine is inversely related to glomerular filtration rate (GFR) and, although an imperfect measure, it is commonly used to assess renal filtration function.

- Reference Range values:
    - Males: 0.9 – 1.3 mg/dL
    - Females: 0.6 – 1.1 mg/dL
- Biochemistry
    - Creatine is converted into creatine phosphate in muscle tissue and is used as a high energy source. It loses phosphate and water to becomes creatinine which diffuses back into plasma and is then excreted in urine
- Clinical Applications
    - Elevated creatinine is associated with abnormal renal function, specifically as it relates to glomerular function. Plasma concentration of creatinine is inversely proportional to the renal clearance of creatinine. Thus, when plasma creatinine concentration is elevated, clearance of creatinine is decreased, indicating renal damage. Plasma creatinine is a relatively insensitive marker and may not be measurably increased until renal function has deteriorated more than fifty percent.
- Analytical Method
    - In the kinetic Jaffe method, serum is mixed with alkaline picrate producing a orange-red complex thus the rate of change in absorbance is measured.
    - Enzymatic coupling reactions
        - If diabetes is suspected

- Pathophysiology
    - Elevated creatinine is associated with abnormal renal function.
    - The ratio BUN/[Creatinine] indicates azotemia

Creatine

In muscle disease such as muscular dystrophy, poliomyelitis, hyperthyroidism, and trauma, both plasma creatine and urinary creatinine are often elevated. Plasma creatinine concentrations are normal in these patients. Plasma creatine concentration is not elevated in renal disease. Creatine is not an ideal biomarker thus measurement of creatine kinase is used typically for the diagnosis of muscle disease because analytic methods for creatine are not readily available in most clinical laboratories.

Uric Acid

Uric acid is the final breakdown product of purine catabolism. It is produced by the liver and is eliminated by the kidney and the gastrointestinal tract. Uric acid in plasma exist as monosodium urate. At the physiological pH, urate is relatively insoluble. At concentrations greater than 7 mg/dL, the plasma is saturated with monosodium urate. As a result, urate crystals may form and precipitate in the tissues. The physiological pH of urine is between 4 – 8. When urine pH is less than 5, uric acid is the predominant species and uric acid crystals may form.

- Reference Range values:
    - Males: 2.5 – 8 mg/dL
    - Females: 1.9 – 7.5 mg/dL
- Biochemistry
    - Product of the catabolism of purine nucleic acids
- Clinical Applications
    - Uric acid is measured to confirm diagnosis and monitor treatment of gout, to prevent uric acid nephropathy during chemotherapeutic treatment, to assess inherited disorders of purine metabolism, to detect kidney dysfunction, and to assist in the diagnosis of renal calculi.
- Analytical Method
    - Coupled enzyme method using uricase and hydrogen peroxidase to produce allantoin
- Pathophysiology
    - Abnormally increased plasma uric acid is found in gout and renal disease
    - Lesch-Nyhan syndrome can result in increased uric acid levels
    - Decreased uric acid levels can be the result of chemotherapy

Ammonia

Ammonia during protein metabolism. Free ammonia is toxic; however, ammonia is present in the plasma in low concentrations.

- Reference Range
    - Plasma: 19 – 60 ug/dL
- Biochemistry
    - Ammonia is produced by the catabolism of amino acids and by bacterial metabolism in the lumen of the intestine. Some endogenous ammonia results from anaerobic metabolic reactions that occur in skeletal muscle during exercise.
- Clinical Applications
    - Blood ammonia concentrations can provide useful clinical information for hepatic failure, Reye's syndrome and deficiency of urea cycle enzymes. Reye's syndrome, occurring most commonly in children, is a serious disease that can be fatal. Frequently, the disease is preceded by a viral infection and the administration of aspirin. Reye's syndrome is an acute metabolic disorder of
- Analytical Methods
    - Direct measurement of ammonia by enzymatic method or ion selective electrode
- Pathophysiology
    - High concentrations of NH_3 are neurotoxic and often associated with encephalopathy

Kidney Function

The kidneys are paired, bean-shaped organs located on either side of the spinal column. Each kidney contains approximately 1 million nephrons. The kidneys filter the blood to remove waste products and toxins and transfer them into the urine for elimination. When the kidneys are not functioning correctly two kinds of problems occur:

- Toxins and NPNs that should be filtered from the blood into the urine are not and are found in high concentrations in the blood. Two examples are creatinine and urea nitrogen (BUN).
- Blood substances that should not be filtered into the urine, but should be held back by the kidney, are escaping into the urine, resulting in high levels in urine and low levels in the blood (ie. albumin and glucose).

The Nephron

A nephron is the basic structural and functional unit of the kidneys that regulates water and soluble substances in the blood by filtering the blood, reabsorbing what is needed, and excreting the rest as urine. Its function is vital for homeostasis of blood volume, blood pressure, and plasma osmolarity.

The glomerulus is the first part of the nephron and functions to filter incoming blood. The glomerulus contains capillary tufts surrounded by the expanded end of a renal tubule known as Bowman's capsule. Each glomerulus is supplied by an afferent arteriole carrying the blood in and an efferent arteriole carrying the blood out. The efferent arteriole branches into capillaries that supply the tubule. The proximal convoluted tubule is located in the cortex. The long loop of Henle is composed of the thin descending limb, which spans the medulla, and the ascending limb, which is located in both the medulla and the cortex, composed of a region that is thin and then thick. The distal convoluted tubule is located in the cortex. The collecting duct is formed by two or more distal convoluted tubules as they pass back down through the cortex and the medulla to collect the urine that drains from each nephron. Collecting ducts eventually merge and empty their contents into the renal pelvis.

The kidney filtration process takes place in structural regions called glomeruli and the filtration is often assessed by a concept called glomerular filtration rate or GFR. The GFR is expressed as the volume of blood plasma that is cleared of a specified substance each minute. A number of different substances can be used to determine GFR, but the most common is creatinine.

Creatinine Clearance

Creatinine clearance requires a carefully timed urine collection to be able to measure the amount of creatinine excreted over a known time period. This is accompanied by a blood sample taken either at the beginning or end of the urine collection period to measure the plasma concentration.

$$GFR = \text{Urine Creatinine} \times \text{Rate of urine output} \left(\frac{mL}{min}\right) [\text{Plasma Creatinine}]$$

If 1 gram of creatinine is excreted into 1.0 liter of urine (1000 mg/L or 100 mg/dL) in a 24-hour period (1440 min) and the plasma creatinine concentration is 0.7 mg/dL the GFR would be:

$$GFR = \frac{100 \frac{mg}{dL} \times \frac{1000 mL}{1440 \min}}{0.7 \frac{mg}{dL}} = 99 \frac{mL}{min}$$

In adults, GFR can range from 50 to 150 mL/min, with younger people having higher values and older people having lower values. The collection of an accurately timed urine sample is often problematic. Poorly collected or poorly timed samples may introduce errors into determination of GFR.

Definition and Key Terms

Blood Urea Nitrogen (BUN)
Test measures the amount of nitrogen in your blood that comes from the waste product urea.

Chronic kidney disease (CKD)
CKD can be broken down into five stages. Physicians, with the help of some key laboratory tests, can determine the appropriate stage for the patient.

STAGE	Description	GFR (ml/min)
S_0	Normal kidney function	91 or above
S_1	Kidney damage (e.g., protein in the urine)	90
S_2	Kidney damage with mild decrease in GFR	60 to 89
S_3	Moderate decrease in GFR	30 to 59
S_4	Severe reduction in GFR	15 to 29
S_5	Kidney failure	Less than 15

Claudication
Pain, commonly in the legs, caused by too little blood flow, usually during exercise. Often indicates peripheral artery disease.

Creatinine
A nonprotein nitrogen compound derived from the spontaneous hydrolysis of creatine or the cyclization of phosphocreatine; creatinine production is relatively constant, is related to muscle mass, and is used as a marker of the glomerular filtration rate of the kidneys.

Gout
A group of disorders of purine metabolism; can be due to primary (inherited) or secondary causes such as chronic kidney disease.

Hyperuricemia
An excess of uric acid or urates in the blood with many causes; it is a precondition for the development of gout and may lead to renal disease.

Hypouricemia
Decreased uric acid concentration in the blood, secondary to a number of underlying conditions such as severe hepatocellular disease and defective renal tubular reabsorption.

Jaffe reaction
The reaction of creatinine with alkaline picrate to form a colored compound; this creatinine assay is subject to numerous interferences.

Urea
The major nitrogen-containing metabolic product of protein catabolism in humans.

Urease methods
Enzymatic assays that initially involve the hydrolysis of urea by urease to generate ammonia, which is quantified by a variety of methods.

Uric acid
A nitrogenous compound derived from the catabolism of purine nucleosides.

Uricase methods
A group of enzymatic assays that initially involve oxidation of uric acid by uricase to eventually produce a compound that is spectrophotometrically measured to determine uric acid concentration.

Non-Protein Nitrogen Problems

1. Describe the major pathological conditions associated with increased and decreased plasma concentrations of urea, uric acid, creatinine, creatine, and ammonia.

2. List the commonly used methods for the determination of urea, uric acid, creatinine, creatine, and ammonia.

3. Consider the anatomy of the nephron. Describe the physiologic role of each part of the nephron: glomerulus, proximal tubule, loop of Henle, distal tubule, and collecting duct

4. What is the significance of glomerular filtration rate and estimated glomerular filtration rate.

5. Define azotemia and uremia.

6. Outline common causes of prerenal, renal, and postrenal azotemia.

7. List the causes of a decreased BUN.

8. A technologist obtains a urea N value of 61 mg/dL and a serum creatinine value of 2.5 mg/dL on a patient. These results indicate what type of azotemia?

9. A patient complaining of lower back pain is seen by a physician. A CLS obtains a BUN value of 19 mg/dL and a serum creatinine value of 2.0 mg/dL on a patient. The results would indicate what condition and why?

10. Calculate the GFR, given the following information below. What stage of CKD is the patient?
    serum creatinine, 1.2 mg/dL
    urine creatinine, 120 mg/dL
    urine volume, 1750 mL/24 hr

11. A CLS obtains a BUN value of 35 mg/dL and a serum creatinine value of 3.2 mg/dL on a patient. The results would indicate what condition and why? What additional test would you run to rule out other possible conditions?

12. An 88-year-old man who had suffered a flu-like illness for the prior three weeks is seen by his primary care doctor. Clinical examination includes urinalysis, which indicates that his urine is strongly positive for protein on dipstick testing. A 24-hour urine collection done to confirm the dipstick finding shows gross proteinuria of 4g/24-hours. The patient has edema of both ankles and his blood pressure is 162/94 mmHg. What additional tests should be run? Explain the rational for each test proposed.

13. A 57-year-old man is seen by his primary care doctor for management of his cardiovascular risk. Modifiable risk factors identified include: the fact that he smokes twenty cigarettes daily and his weight of 345 lbs. He is also hypertensive with a blood pressure reading of 152/89 mmHg. On systems review he admits to symptoms suggestive of intermittent claudication. He is counselled about smoking, given dietary advice, and an ACE inhibitor is prescribed. What additional test should be run? Explain the rational for each test proposed. How would his smoking confound test results?

Non-Protein Nitrogen Compounds

14. A 45-year-old male presents to the emergency department complaining of intense joint pain. The previous night the patient experienced similar pain accompanied by inflammation and redness of his wrists and large toe. What additional test should be run? Explain the rational for each test proposed.

15. A 65-year-old man was first admitted for treatment of chronic obstructive lung disease, renal insufficiency, and significant cardiomegaly. What additional test should be run? Explain the rational for each test proposed.

Matching

16. The compound that is not an NPN _____
    The major excretory product _____
    BUN is measured using this method _____
    BUN:[Cr]p = 25 _____
    Elevated plasma Uric acid is associated with this _____
    Creatinine plus what reagent creates a red-orange complex _____
    Toxic effects of elevated blood ammonia cause _____
    A medical condition characterized by abnormally high Urea levels _____
    BUN:[Cr]p < 10 _____
    Plasma creatinine levels are inversely related to _____
    a. Urea
    b. Decrease mental status
    c. Spectroscopy
    d. Jaffe method
    e. Coupled Enzymatic methods
    f. Congestive Heart Failure
    g. Allantoin
    h. Renal Failure
    i. GFR
    j. Azotemia
    k. Aminoacidopathy
    l. Lesch-Nyhan Syndrome
    m. Alkaline Picric acid

    Answer:
    g
    a
    e
    f
    l
    m
    b
    j
    h
    i

CHAPTER 13
Enzymes

Clinical Significance
Enzymes are proteins with catalytic activity; they accelerate the rate at which a chemical reaction takes place without themselves being consumed in the process. Different forms of enzymes exist that are referred to as isoenzymes. Enzymes are the ideal markers in various disease states such as liver disease, myocardial infarction, jaundice, pancreatitis, cancer, neurodegenerative disorders, heart disease, renal disease as well as alcoholism. Enzymes play an integral role in many cellular processes. Thousands have been identified however the clinical laboratory routinely measures fewer than 15 for diagnostic purposes. This chapter will review the commonly analyzed enzymes, including their tissue source and clinical significance (35).

Liver Biomarkers
This section is devoted to clinical tests that are done to assess how well the liver is functioning. These test measure certain proteins and enzymes that are normally released by liver cells and are usually performed in patients who are known to have or are suspected of having liver disease, clinical signs of jaundice, hepatitis and chronic alcoholism (29, 30, 36, 37).

Alanine Aminotransferase (ALT)
Reference Range: 7–45 U/L
An increased ALT level is often a sign of liver disease. ALT catalyzes the transfer of an amino group from alanine to α-ketoglutarate via the formation of glutamate and pyruvate. ALT is distributed in many tissues, with comparatively high concentrations in the liver. It is considered the more liver-specific enzyme of the transferases. An increased ALT level may be due to any of the following:

- Cirrhosis of the liver
- Liver necrosis
- Hepatitis
- Hemochromatosis
- Fatty liver
- Liver ischemia
- Liver tumor or cancer
- Use of drugs that are toxic to the liver
- Mononucleosis
- Pancreatitis

Aspartate Aminotransferase (AST)
Reference Range: 7–42 U/L
This enzyme is largely present in the liver and heart, but may also be seen in the kidney, brain, muscles, and red blood cells. If any of these structures are damaged there is an increase in AST.

Decreased values of AST may be seen in congested liver, or in patients with high cholesterol levels however an increased in AST may be due to various causes as follows: Liver diseases, alcoholism, Myocardial infarction Kidney infections and diseases AST is an indicator thus a rise in the level of AST must always be taken in combination with ALT levels. If ALT levels are normal, then, AST indicates damage to some other organ, namely cardiac muscle.

The DeRitis ratio, AST:ALT, can be used to detect alcohol-induced liver disease. An AST:ALT value of greater than 2 indicates this condition. An AST:ALT value less than or equal to 1 indicates acute viral hepatitis.

Alkaline Phosphatase (ALP)
Reference Range: 20-140 IU/L
Alkaline phosphatase is an enzyme that is present not only in the liver, but also in the bone and intestine. This enzyme had many metabolic functions, and is believed to play an important role in skeletal mineralization.

Increased levels of alkaline phosphatase may be found in the following conditions:
- Liver diseases – Cirrhosis, hepatitis, biliary obstruction
- Bone diseases – Bone tumors, osteoporosis, rickets, Paget's disease
- hyperparathyroidism
- leukemias and lymphomas

Decreased levels of alkaline phosphatase may be seen in:
- Wilson's disease
- pernicious anemia
- aplastic anemia
- chronic myelogenous leukemia

5'-Nucleotodase (5NT)
Although 5NT is found in a wide variety of cells, serum levels become significantly elevated in hepatobiliary disease. There is no bone source of 5NT, so it is useful in differentiating ALP elevations due to the liver from other conditions where ALP may be seen in increased concentrations.

Gamma Glutamyl Transferase (GGT)
Reference range: 7-30 IU/L
This enzyme is found in the liver, kidney, and pancreas. Although it is present in higher amounts in the kidney, its activity is greatest in the liver.

Increased levels of GGT are usually associated with the following conditions:
- Liver diseases – cirrhosis, hepatitis, alcoholic liver disease, liver cancer
- Cancers of the prostate, breast, and lung
- Pancreatitis
- Systemic lupus erythematosus

GGT levels may be decreased in:
- hyperthyroidism
- hypothalamic dysfunction
- low magnesium levels.

Similar to 5NT, GGT is an indicator enzyme. GGT levels may be used along with elevated ALP levels to establish the disease source. If GGT and ALP are both elevated, it is indicative of hepatobiliary disease, whereas, an isolated increase in ALP would suggests bone disease.

Lactate Dehydrogenase (LD)
Reference Range: 20-250 IU/L
LD is an enzyme with a very wide distribution throughout the body. It consists of 6 isoenzymes ranging from LD-1 to LD-6. Lactate dehydrogenase is released into circulation when cells are damaged or destroyed, serving as a general, nonspecific marker of cellular injury. High serum levels of LD-5 may be found in metastatic liver carcinoma.

Protein Liver Biomarkers

Serum Bilirubin
Although bilirubin is not an enzyme, its concentration is used alongside enzymatic test to determine liver disfunction. Bilirubin is a breakdown product of hemoglobin. Its unconjugated form passes to the liver, where it is conjugated and subsequently excreted largely through the urine. Both the conjugated, direct, and unconjugated, indirect, form can be measured from serum.

- Bilirubin Reference Range
    - Total bilirubin: 0.2 – 1.2 mg/dl
    - Direct bilirubin: 0.1 – 0.4 mg/dl
    - Indirect bilirubin: 0.4– 1.1 mg/dl

Elevated bilirubin levels are indicative of jaundice, which could be due to several reasons. Pre-hepatic Jaundice is caused by an increase in indirect bilirubin. Pre-hepatic jaundice could be due to Increased hemolysis from: malaria, sickle cell anemia, thalassemia, hemolytic anemia. Hepatic Jaundice is an increase in both direct and indirect bilirubin. Symptoms are dark urine and pale stools and could be due to, hepatitis, cirrhosis or alcoholic liver disease. Post hepatic Jaundice is an increase in direct bilirubin which results in, gallstones, bile duct strictures, biliary atresia and pancreatic cancer (38).

Cardiac Biomarkers
One of the most common medical emergencies that can occur with regard to the cardiac system is a myocardial infarction. Infarction is commonly secondary to decreased blood supply to the myocardium of the heart, which causes ischemia which causes irreversible necrosis of myocytes, which is referred to as an infarct. When the myocardium gets damaged, certain enzymes get released into the blood stream in increased levels. These are useful clinically as biomarkers to diagnose myocardial infarction. **The primary enzyme and protein used as cardiac markers are given below:**

Cardiac Enzymes
- Creatine kinase
- Lactate dehydrogenase
- AST

Cardiac Protein Markers
- Cardiac troponins
- Myoglobin

Creatine Kinase
Reference Range: 20-200 U/L

In heart, skeletal muscle, and brain cells, creatine phosphokinase (CK) catalyzes the conversion of creatine to creatinine. Because this is a catabolic process, the enzyme's levels rise as a result of tissue injury. CK has three isozymes, CK-MM, CK-MB, and CK-BB.

Brain tissue has CK-BB, whereas skeletal and cardiac muscle contains CK-MM and CK-MB. CK-MB is the most susceptible to myocardial injury of the three. It rises 4–6 hours after an acute MI, peaks in 18–24 hours, and then falls back to normal in 3–4 days. Because this enzyme's elevation lasts only a few minutes, it can't be used to make a late diagnosis of acute MI. However, if the numbers rise again, it could indicate infarct extension. CK-MB normal range: 0–4 ng/ml.

Lactate Dehydrogenase (LD)
Reference Range: 20-250 IU/L

Like the liver, the heart uses LD as a biomarker. Isoforms, LD-1 and LD-2, are found in the heart, red blood cells, and kidneys, respectively. After a MI, the total LD level rises 2–5 days. For ten days, the increased level persists. Several factors can cause total LD to rise. The amounts of isozymes can sometimes be altered, even though the total value is within normal ranges. As a result, it's crucial to keep track of the various isozyme ratios. LD-1 accounts for 17.5 percent to 28.3 percent of the total, whereas LD-2 accounts for 30.4 percent to 36.4 percent. In most cases, LD-1 levels are lower than LD-2. The level of LD-2 does not alter after an acute MI, while the level of LD-1 increase. This is referred to as a reversed state commonly known as the LD-1>LD-2 flip pattern. After a MI, this inverted pattern develops 12-24 hours later and lasts for 48 hours.

Aspartate Aminotransferase (AST)
Reference Range: 7–42 U/L

After myocardial infarction, AST, which is mostly found in the liver and heart, tends to rise 8 to 12 hours later. It peaks between 24 and 48 hours following an infarction. It is not specific to MI because serum levels of this enzyme might rise in other illnesses such as liver disease, pancreatitis, and so on. For a more definitive diagnosis of MI, it is utilized in conjunction with other enzyme results.

Protein Cardiac Biomarkers

Troponins
Reference Range: 0.01 ng/ml

Troponin is a type of protein found in your heart's muscles. Troponin is a protein that isn't ordinarily seen in the bloodstream. Troponin is released into the bloodstream when cardiac muscles are injured. Troponins can also be elevated in cases of pulmonary embolism, myocarditis, and cardiac failure. As a result, laboratory results must be interpreted in conjunction with clinical findings. As cardiac damage worsens, more troponin is released. Troponin levels in the blood may indicate that you are experiencing or have recently had a heart attack. The three subunits of this enzyme are troponin I, troponin-C, and troponin-T. The only troponins that are specific to the myocardium are troponin-I and troponin-T. Troponin is measured in micrograms per milliliter and has a normal value of 0.01 ng/ml. After MI, this enzyme is released for 2–4 hours before peaking at 10–24 hours. Troponins can also be utilized to determine the size of an infarct. For this, the values must be measured on the third day (39, 40).

Myoglobin
Reference Range: 30–90 ng/ml
Myoglobin is a protein that has a similar structure to hemoglobin. It is located in muscle tissue and is responsible of iron and oxygen binding and 30–90 ng/ml is considered normal. Myoglobin levels are used in conjunction with cardiac enzymes to confirm myocardial injury, despite the fact that it is technically not a cardiac protein seen in myocytes. Myoglobin aids in the estimation of muscle damage but does not reveal the location of injury. Within 1 - 2 hours of a MI, this protein becomes abnormal, with a peak tome of 4 - 8 hours. In 12 to 24 hours levels return to normal.

Pancreases

Amylase (AMY)
Reference Range: 28–100 U/L
Amylase (AMY) belongs to the hydrolase family of enzymes that catalyze the breakdown of starch and glycogen. There are two isoforms of amylase. P- isoamylase is derived from pancreatic tissue while S-isoamylase is derived from salivary gland tissue, as well as the fallopian tube and lung. Once the polysaccharides reach the intestine, pancreatic AMY performs the primary digestive action of starches.

In the diagnosis of acute pancreatitis, serum and urine AMY values have diagnostic value. Serum AMY levels rise 5 to 8 hours after the commencement of an incident in acute pancreatitis, peak at 24 hours, and recover to normal within 3 to 5 days. Salivary gland lesions, such as mumps and parotitis, and various intra-abdominal diseases, such as perforated peptic ulcer, intestinal obstruction, cholecystitis, and ruptured ectopic pregnancy, can all cause an increased serum AMY level.

Lipase (LPS)
Reference Range: 0 – 30 U/L
Lipase is mostly found in the pancreas; however, it can also be found in the stomach and small intestine. Serum LPS activity rises 4 to 8 hours after an acute pancreatitis episode, peaks at 24 hours, and then declines over the next 8 to 14 days. LPS measurements are similar to AMY readings in this regard, however LPS is thought to be more specific for pancreatic diseases. Both AMY and LPS levels rise immediately, while LPS elevations in acute pancreatitis last until around 8 days.

Enzymes

Definitions and Key Terms

Activator
Inorganic cofactors: ex chloride or magnesium ions

α-Amylase
An enzyme that catalyzes the hydrolysis of 1,4-alpha-glycosidic linkages in starch, glycogen, and related polysaccharides and oligosaccharides.

Acid phosphatase
All phosphatases with optimal activity below pH 7.0 that catalyze the cleavage of orthophosphate from orthophosphoric monoesters; most of the activity in serum is of a tartrate-resistant type.

Acute pancreatitis
A sudden inflammation of the pancreas that is usually accompanied with severe upper abdominal pain.

Aldolase
A lyase that catalyzes cleavage of fructose-1,6-diphosphate into dihydroxyacetone-phosphate and glyceraldehyde 3-phosphate in the glycolytic breakdown of glucose to pyruvate.

Alkaline phosphatase
A hydrolase that catalyzes the alkaline hydrolysis of a large variety of naturally occurring and synthetic substrates.

Aminotransferases
A subclass of enzymes of the transferase class that catalyze the transfer of an amino group from a donor (generally an amino acid) to an acceptor (generally a 2-oxo acid). Most of these enzymes are pyridoxal phosphate proteins. Alanine and aspartate aminotransferase are examples that are of significant clinical utility.

Apoenzyme
The protein component of an enzyme. Cholecystitis A painful inflammation of the gallbladder.

Cholinesterase
An enzyme of the hydrolase class that catalyzes the cleavage of the acyl group from various esters of choline, including acetylcholine, and some related compounds.

Coenzyme
An organic nonprotein molecule that binds with the protein molecule (apoenzyme) to form the active enzyme (holoenzyme).

Cofactor
A molecule which may be necessary for enzyme activity

Creatine kinase
A dimeric transferase enzyme that catalyzes the reversible phosphorylation of creatine by adenosine triphosphate (ATP). Creatine kinase (CK) has four isoenzymes: CK-MM, CK-MB, CK-BB, and mitochondrial CK.

γ-Glutamyltransferase
A transferase enzyme that reversibly catalyzes the transfer of a glutamyl group from a glutamyl-peptide and an amino acid to a peptide and a glutamyl-amino acid.

Holoenzyme
Active enzyme formed by combination of a coenzyme and an apoenzyme.

Isoenzyme
A molecular form that originates at the level of the genes that encode the structures of the enzyme proteins in question.

Isoform
An enzyme molecular form that has been post-translationally modified.

Lactate dehydrogenase
An oxidoreductase enzyme (L)-lactate, using NADH (reduced form of nicotinamide adenine dinucleotide) as an electron donor.

Lipase
A hydrolase that hydrolyzes glycerol esters of long-chain fatty acid.

5'-Nucleotidase
A phosphatase that acts only on nucleoside-5'-phosphates, such as adenosine-5'-phosphate (AMP), releasing inorganic phosphate.

Paget disease
A chronic disorder that results in enlarged and deformed bones (also known as osteitis deformans, osteodystrophia deformans).

Prosthetic group
A tightly bound, nonpeptide structure required for the activity of an enzyme.

Enzyme Problems

1. The DeRitis ratio (AST:ALT) less than or equal to 1 is associated with which condition?
   a. Acute viral hepatitis
   b. Cirrhosis
   c. Alcoholic hepatitis
   d. All of the above
   Answer: a

2. The DeRitis ratio (AST:ALT) greater than 2 is associated with which condition?
   a. Acute viral hepatitis
   b. Cirrhosis
   c. Alcohol induce liver disease
   d. All of the above
   Answer: c

3. The most sensitive indicator of alcoholic liver disease is:
   a. AST.
   b. ALT.
   c. GGT.
   d. ALP.
   Answer: c

4. AST is elevated in diseases involving the:
   a. Renal cortex.
   b. Heart muscle.
   c. Intestinal epithelium.
   d. Acinar cells of the pancreas.
   Answer: b

5. A substrate used in lipase reactions is:
   a. Starch.
   b. ρ-nitrophenyl phosphate.
   c. Olive oil emulsion.
   d. Creatine phosphate.
   Answer: c

6. The best test for pancreatitis is:
   a. Virus isolation.
   b. A serologic test for mumps.
   c. A urine amylase test.
   d. A serum lipase test.
   Answer: d

Enzymes Case Studies

7. Serum chemistry results on a 53-year-old female are

   AST: 120 U/L (4-34U/L)
   ALT: 185 U/L (5-40 U/L)
   ALP: 785 U/L (<165 U/L)
   GGT: 225 U/L (5-24 U/L)
   Total Bilirubin: 10.8 mg/dL
   Direct Bilirubin: 8.6 mg/dL
   urine bilirubin: positive
   urine urobilinogen: Normal
   fecal urobilinogen: decreased

   These results are consistent with a diagnosis of:
   a. Hemolytic anemia.
   b. Hepatitis.
   c. Dubin-Johnson.
   d. Biliary obstruction.
   Answer: d

8. Mrs. Smith has the following chemistry results:

   CK: moderately increased
   LD: moderately increased
   LD-1: increased
   LD-2: moderately increased

   The probable diagnosis is:
   a. Myocardial infarction.
   b. Acute pancreatitis.
   c. Acute hepatitis.
   d. Cholecystitis.
   Answer: a

9. Although a total lactate dehydrogenase (LD) determination performed alone yields little information as to the area of tissue destruction, an electrophoretic separation of LD isoenzyme may be useful. With what disorder is an increase in LD-5 and LD-4 associated?
   a. Acute hepatic disease
   b. Acute myocardial infarction
   c. Pulmonary infarction
   d. Pancreatitis
   Answer: a

CHAPTER 14
Carbohydrate

Clinical Significance
Diabetes is a group of conditions linked by an inability to produce enough insulin and/or to respond to insulin. This causes high blood glucose levels (hyperglycemia) and can lead to a number of acute and chronic health problems, some of them life-threatening.

Diabetes is the seventh leading cause of death in the United States. According to the Centers for Disease Control and Prevention, about 29 million people in the U.S. currently have diabetes, but as many as 8 million are not yet aware that diabetes is affecting their health.

People with diabetes are unable to process glucose, the body's primary energy source, effectively. Normally, after a meal, carbohydrates are broken down into glucose and other simple sugars. This causes blood glucose levels to rise and stimulates the pancreas to release insulin into the bloodstream. Insulin is a hormone produced by the beta cells in the pancreas. It regulates the transport of glucose into most of the body's cells and works with glucagon, another pancreatic hormone, to maintain blood glucose levels within a narrow range.

If someone is unable to produce enough insulin, or if the body's cells are resistant to its effects (insulin resistance), then less glucose is transported from the blood into cells. Blood glucose levels remain high, but the body's cells starve thus cannot produce ATP. This can cause both short-term and long-term health problems, depending on the severity of the insulin deficiency and/or resistance. Diabetics typically have to control their blood glucose levels on a daily basis and over time to avoid health problems and complications. Treatment, which may involve specialized diets, exercise and/or medications, including insulin, aims to ensure that blood glucose does not get too high or too low (41-45).

Hyperglycemia
A very high blood glucose level (acute hyperglycemia) can be a medical emergency. The body tries to rid the blood of excess glucose by flushing it out of the system with increased urination. This process can cause dehydration and upset the body's electrolyte balance as sodium and potassium are lost in the urine. With severe insulin deficiency, glucose is not available to the cells and the body may attempt to provide an alternate energy source by metabolizing fatty acids. This less efficient process leads to a buildup of ketones and upsets the body's acid-base balance, producing a state known as ketoacidosis. Left unchecked, acute hyperglycemia can lead to severe dehydration, loss of consciousness, and even death.

Hypoglycemia
A very low blood glucose level (hypoglycemia), often as a result of too much insulin, can also be life-threatening. It can lead to hunger, sweating, irregular and rapid heartbeat, confusion, blurred vision, dizziness, fainting, and seizures. Severely low blood glucose can lead fairly quickly to insulin shock and death.

Glucose levels that rise over time and become chronically elevated may not be initially noticed. The body tries to control the amount of glucose in the blood by increasing insulin production and by eliminating glucose in the urine. Signs and symptoms usually begin to arise when the body is no longer able to compensate for the higher levels of blood glucose.

Chronic high blood glucose can cause long-term damage to blood vessels, nerves, and organs throughout the body and can lead to other conditions such as kidney disease, loss of vision, strokes, cardiovascular disease, and circulatory problems in the legs. Damage from hyperglycemia is cumulative and may begin before a person is aware that he or she has diabetes. The sooner that the condition is detected and treated, the better the chances are of minimizing long-term complications.

Type of Diabetes
Type 1
Exact cause unknown; thought to be primarily an autoimmune disease that involves the destruction of the insulin-producing beta cells in the pancreas; can occur at any age but usually diagnosed in children and young adults. Individuals show large amounts of ketones in urine.

Type 2
Most common type; associated with insulin resistance and with insulin production that is insufficient to meet the body's needs and to compensate for resistance. It develops most frequently in overweight middle-aged and elderly people. With increased obesity in children and adolescents, the condition is becoming more common at younger ages.

Gestational
Develops during a woman's pregnancy and affects both mother and developing baby; typically develops late in the pregnancy.

Prediabetes
Higher blood glucose than normal, but not considered diabetes; people with prediabetes are at an increased risk of developing diabetes.

Signs and Symptoms
The signs and symptoms of diabetes are related to hyperglycemia, hypoglycemia, and complications associated with diabetes. Type 1 diabetics are often diagnosed with acute severe symptoms that require hospitalization. With prediabetes, early type 2 diabetes, and gestational diabetes, there usually are no signs or symptoms. Type 1 and type 2 diabetes with hyperglycemia may include:

- Increased thirst
- Increased urination
- Increased appetite (with type 1, weight loss is also seen)
- Fatigue
- Nausea, vomiting, abdominal pain (especially in children)
- Blurred vision
- Slow-healing wounds or infections
- Numbness, tingling, and pain in the feet (neuropathy)
- Erectile dysfunction in men

- Absence of menstruation in women
- Rapid breathing (acute)
- Decreased consciousness, coma (acute)

Temporary hypoglycemia in the diabetic may be caused by the accidental injection of too much insulin, not eating enough or waiting too long to eat, exercising strenuously, or by the swings in glucose levels seen with "brittle" diabetes. Hypoglycemia needs to be addressed as soon as it is noticed as it can rapidly progress to unconsciousness. Signs and symptoms include:
- Sudden severe hunger
- Headache
- Anxiety, confusion
- Sweating
- Trembling, weakness
- Double vision
- Convulsions
- Coma

Screening and Diagnostic Tests

In early stages, diabetes has no obvious symptoms. Glucose screening tests are needed to identify high glucose concentrations in otherwise asymptomatic people. These tests may be done using a fingerstick sample and portable device, as is typical at health fairs, or using a blood sample drawn by venipuncture and measured in a laboratory.

The guidelines for interpretation of screening tests for blood glucose are shown in the table below. Diagnosis of diabetes is based on fasting blood glucose. Sometimes screening may be done when the person is not fasting. In such cases, interpretation is if difficult; however, a non-fasting blood glucose above 200 mg/dL (11.2 mmol/L) is considered to be consistent with diabetes. Impaired glucose tolerance is defined as two-hour glucose levels of 140 to 199 mg/dL (7.8 to 11.0 mmol) on the 75-g oral glucose tolerance test, and impaired fasting glucose is defined as glucose levels of 100 to 125 mg/dL (5.6 to 6.9 mmol/L) in fasting patients.

Fasting Blood Glucose Reference Ranges

70 to 99 mg/dL (3.9 to 5.5 mmol/L)	Non Diabetic
100 to 125 mg/dL (5.6 to 6.9 mmol/L)	Pre Diabetes
126 mg/dL (7.0 mmol/L) and above	Diabetes

Monitoring of Diabetes and Diabetic Complications

A patient with diabetes is at risk for developing a number of complications that are a direct consequence of high levels of glucose. These include kidney failure, blindness, poor circulation leading to foot ulcers, and an increased risk for atherosclerosis and heart disease.

Treatment of diabetes with diet, drugs and insulin is aimed at maintaining blood glucose levels as close to non-diabetic levels as possible. Research studies, including the Diabetes Complication and Control Trial (DCCT), have demonstrated that good control of blood glucose can slow or prevent the development of the many complications that accompany poorly controlled blood glucose.

Diabetic patients often monitor their own blood glucose level regularly to ensure that their diet and medications are appropriate to keep their blood glucose within a target range set by their doctor.

Two other very important tests that are used to con rm glucose control and ensure good kidney function are Hemoglobin A1c and Microalbumin.

Hemoglobin A1c (HbA1c)

Hemoglobin A1c (HbA1c) is a chemically modified hemoglobin molecule. It forms when glucose from the blood enters the red blood cells and attaches to hemoglobin. As glucose concentration in blood increases, more glucose reacts with hemoglobin. Since red blood cells circulate with a half-life of three months, meaning that half of the red blood cells are destroyed and replaced by new ones every three months, the extent to which hemoglobin has been altered by glucose reflects glucose control over the previous three months. Hemoglobin A1c is expressed as a percent that reflects the percentage of hemoglobin molecules that have a glucose molecule attached. There is an increasing interest in using HbA1c to screen for diabetes.

Estimated Average Glucose (eAG)

Actual blood glucose varies widely over the course of a day, rising after meals and falling gradually to fasting levels. HbA1c reflects the average blood glucose concentration and glycemic control. There are formulas available that allow eAG to be calculated from HbA1c values.

Urine Albumin

Urine albumin (microalbumin) is a test for very small amounts of albumin escaping from the kidney and leaking into the urine. The first sign of albumin in the urine is a signal that kidney function is being compromised. Early detection can lead to more aggressive treatment to prevent continuing damage to the kidney.

Definitions and Key Terms

Aldehyde
An organic compound with a carbonyl group (a carbon atom double-bonded to an oxygen) at the end of the carbon chain bonded to hydrogen and an R group (usually an alkyl group).

Carbohydrate
Aldehyde or ketone derivatives of polyhydroxy alcohols composed of carbon, hydrogen, and oxygen in a ratio of 1:2:1.

Diabetes mellitus
A group of metabolic disorders of carbohydrate metabolism in which glucose is underutilized, producing hyperglycemia.

Glucagon
A protein hormone that maintains blood glucose concentration by increasing blood glucose through glycogenolysis.

Glucose
A six-carbon monosaccharide derived from the breakdown of carbohydrates in the diet or in body stores; can also be endogenously synthesized from protein or the glycerol moiety of triglycerides.

Glycogen
An extensively branched polysaccharide containing many glucose residues and found particularly in muscle and liver cells for glucose storage.

Hypoglycemia
Blood glucose concentration in the blood decreased below a healthy reference interval.

Insulin
A protein hormone that maintains blood glucose concentration by decreasing blood glucose through cellular uptake.

Ketone
An aldehyde that has a carbonyl group (carbon atom double-bonded to an oxygen atom) at any position other than at the end of the carbon chain.

Lactate
An intermediary product in glucose metabolism that accumulates in the blood predominantly when tissue oxygenation is decreased, as during strenuous exercise; an increased blood lactate concentration is called lactic acidosis.

Pyruvate
An organic acid formed from glucose through glycolysis.

Carbohydrate Problems

1. Explain the changes that occur in the body with hyperglycemia.

2. Define hypoglycemia and discuss the common causes of drug-induced, reactive, and fasting hypoglycemia.

3. What are the three factors in Whipple's triad.

4. An isomer of glucose with the -OH group of the anomeric carbon C1 that is below the plane of the ring or on the right-hand side is:
    a. D-glucose.
    b. L-glucose.
    c. α -glucose.
    d. β -glucose.
    Answer: c

5. Which of the following carbohydrates is a polysaccharide?
    a. Starch
    b. Sucrose
    c. Lactose
    d. Glucose
    Answer: a

6. Gluconeogenesis is:
    a. The conversion of glucose to glycogen for storage.
    b. The formation of glucose from noncarbohydrate sources, for example, amino acids, glycerol, and lactate.
    c. The conversion of glucose into 3-C molecules, for example, lactate and pyruvate.
    d. Breakdown of glycogen to form glucose and other intermediate products.
    Answer: b

7. Glycolysis is:
    a. The conversion of glucose into lactate or pyruvate and then CO_2 and H_2O.
    b. The conversion of glucose to glycogen for storage.
    c. The breakdown of glycogen to form glucose and other intermediate products.
    d. The formation of glucose from noncarbohydrate sources.
    Answer: a

8. Glycogen is stored in the:
    a. Pancreas.
    b. Liver.
    c. Spleen.
    d. Gall bladder.
    Answer: b

Carbohydrate

9. Which of the following is the primary hypoglycemic hormone?
    a. Insulin
    b. Thyroxine
    c. Glucagon
    d. Growth hormone
    Answer: c

10. Which of the following hormones does NOT stimulate glycogenolysis?
    a. Insulin
    b. Glucagon
    c. Epinephrine
    d. Thyroxine
    Answer: a

11. The only hormone that causes a decrease in blood glucose levels is:
    a. Glucagon.
    b. Thyroxine.
    c. Insulin.
    d. Parathyroid hormone.
    Answer: c

12. Which of the following is characteristic of type 2 diabetes mellitus?
    a. High insulin levels
    b. Ketosis
    c. Obesity and physical inactivity
    d. Juvenile onset
    Answer: c

13. Which form of diabetes usually manifests itself early in life, and is associated with ketosis, low insulin levels, and autoantibodies to islet cells?
    a. Gestational
    b. Type 1
    c. Type 2
    Answer: b

14. All of the following are associated with gestational diabetes EXCEPT:
    a. It converts to diabetes mellitus after pregnancy in 30-60% of patients.
    b. Is diagnosed using the same glucose tolerance criteria as in nonpregnant women.
    c. Is defined as glucose intolerance during pregnancy.
    d. Associated with increased fetal risk.
    e. Answer: b

15. All of the following are confirmatory of diabetes mellitus EXCEPT:
    a. Fasting glucose greater than 126 mg/dL.
    b. Urine glucose greater than 300 mg/dL
    c. 2-hour postprandial glucose greater than 200 mg/dl
    d. 1-hour and 2-hour glucose tolerance values greater than 200 mg/dL.
    Answer: b

16. Complications of diabetes mellitus include all of the following EXCEPT:
    a. Heart disease and stroke.
    b. Neuropathy.
    c. Nephropathy.
    d. Hepatitis.
    Answer: d

17. Instructions for patients preparing for a glucose tolerance tests include all of the following EXCEPT:
    a. No food 10 hours before and during the test.
    b. Patient must be ambulatory for 3 days prior to the test.
    c. Carbohydrate intake must be at least 150g/day for 3 days prior to the test.
    d. Caffeine and smoking are permitted before and during the test.
    Answer: d

18. A patient with an insulinoma may exhibit dizziness and fainting attributable to:
    a. Hypoglycemia.
    b. Hyperglycemia.
    c. Ketosis.
    d. Acidosis.
    Answer: a

19. What type of hypoglycemia is exhibited 8 hours after a meal?
    a. Reactive
    b. Fasting
    c. Alimentary
    d. None of the above
    Answer: b

20. Which of the following fasting 2-hour glucose tolerance results would be classified as impaired glucose tolerance?
    a. 120 mg/dL
    b. 130 mg/dL
    c. 160 mg/dL
    d. 110 mg/dL
    Answer: c

21. A sneaky diabetic tried to lower her glucose by working out and watching her diet 1 or 2 days before her appointment. The rest of the time she spent the day in front of the television and eating chocolates. What test could the doctor order to detect this type of behavior?
    a. A glucose tolerance test
    b. A glycosylated hemoglobin
    c. Daily urine glucose
    d. A hospital admission so he could watch her daily
    Answer: b

CHAPTER 15
Lipids and Lipoproteins

Clinical Significance

Lipids and lipoproteins are essential energy from structural molecules. The energy required for numerous living processes is provided by their breakdown. Some of them, together with proteins, form the most significant structural constituents of cells and cellular organelles, while others serve as precursors to the production of a variety of active substances such as hormones and prostaglandins. Lipids are consumed, but they can also be made in the body. Because lipids are water insoluble, they are found packed in lipoprotein molecules in circulation.

Lipoproteins have a central lipid portion (nucleus) that contains triglycerides and cholesterol esters, and a sheath made up of specific proteins (apoproteins), phospholipids, and small amounts of free cholesterol on the surface. Lipids can be carried into the bloodstream thanks to this sheath. The relevance and role of lipids in the body, as well as their role in many metabolic illnesses and diseases, took a long time to discover.

Abnormalities in lipids and lipoproteins are referred to as hyperlipidemias and hyperlipoproteinemias, respectively. Elevated levels of lipids and lipoproteins are associated with risk for coronary heart disease, specifically, increased total cholesterol, increased LDL, decreased HDL. Pathologic processes are classified based on lipid levels, lipoprotein pattern, and clinical and biochemical phenotype, with most very rare in occurrence (46, 47).

Lipoprotein Differentiation

Chylomicrons are the largest and least dense of lipoprotein molecules, their principal role is the delivery of dietary lipids to hepatic and peripheral cells. Very low-density lipoproteins are produced by the liver and they transport (carry) triglycerides to the peripheral tissues for energy utilization and storage, excess carbohydrate intake results in increased VLDL production. Intermediate Density Lipoproteins exist only in transition of VLDL to LDL. Low density lipoproteins (LDL) are more cholesterol rich than other lipoproteins and are significantly smaller, they are considered to be a marker for coronary heart disease (CHD.) The Friedewald equation is used to estimate LDL as: total cholesterol (TC) minus high-density lipoprotein-cholesterol (HDL) minus triglycerides (TG)/5, with the latter term serving as an estimate for very low-density lipoprotein-cholesterol (VLDL).

Lipoprotein(a) is like LDL and contain one molecule of apo linked to apo B, also elevated levels (>30mg/dL) are thought to increase risk of premature CHD. High density lipoproteins (HDL) are the smallest and densest lipoprotein particles and they can remove excess cholesterol from cells and is said to have an antiatherogenic property. Finally, lipoprotein X is an abnormal lipoprotein, mainly found in patients with cirrhosis or cholestasis.

Apoproteins are functional and structural protein components of the lipoprotein molecule. Each lipoprotein molecule contains one or more apoproteins. There are three main functions: activate enzymes to aid in lipid metabolism, maintain structural integrity of the lipoprotein, and heighten cellular uptake of lipoproteins.

Reference Range
- Total cholesterol:
    - Less than 200 mg/dL
- Triglycerides:
    - less than 150 mg/dL
- HDL
    - Greater than 40 mg/dL;
    - Greater than 60 mg/dL protective (subtract a risk factor)
- LDL
    - Less than 100 mg/dL optimal or:
    - With CHD risk equivalent less than 130 mg/dL with two or more risk factors
    - Less than 160 mg/dL with 0 to 1 risk factor
    - Risk factors
        - Cigarette smoking, hypertension
        - Blood pressure greater than 140/90 mm Hg
        - HDL less than 40 mg/dL,
        - A family history of premature CHD

Definitions and Key Terms

Apolipoproteins
The major protein components of lipoproteins.

Atherosclerosis
A pathogenic process that is the underlying cause of the common cardiovascular disorders of myocardial infarction, cerebrovascular disease, and peripheral vascular disease.

Coronary heart disease
A narrowing of the small blood vessels caused by atherosclerotic plaque that impairs the supply of blood and oxygen to the heart.

Lipids
A class of compounds that are soluble in organic solvents but are nearly insoluble in water and contain nonpolar carbon-hydrogen bonds.

Lipoprotein(a)
A lipoprotein structurally similar to low-density lipoprotein but containing a plasminogen-like protein called apo(a). It carries only a relatively small fraction of total cholesterol but is considered to be particularly proatherogenic.

Lipoproteins
Spherical micelle-like particles involved in the transport of lipids with nonpolar neutral lipids (triglycerides and cholesterol esters) in their core and more polar amphipathic lipids (phospholipids and free cholesterol) at their surface.

Phospholipid
A polar amphipathic lipid located on the surface of a lipoprotein; phospholipids are also found at the aqueous interface of biologic membranes.

Prostaglandin
Potent signaling molecules derived from unsaturated 20-carbon fatty acids, primarily arachidonic acid, by way of the cyclooxygenase pathway. These compounds are involved in a wide variety of physiological processes.

Triglyceride
A glycerol ester consisting of three molecules of fatty acid esterified to glycerol and constituting 95% of tissue storage fat.

Lipids and Lipoproteins Problems

1. Which of the following formulas shows the correct calculation for indirectly measuring LDL-C (Friedewald formula)?
    a. LDL-C = HDL+ (Triglyceride/5)
    b. LDL-C = Total cholesterol − (HDL) − (Triglyceride/5)
    c. LDL-C = Total cholesterol + HDL-C + (Triglyceride/5)
    d. LDL-C = HDL-C − (Triglyceride/5)
   Answer: b

2. Identify the following as HDL, LDL, VLDL, or Chylomicrons.
    - Associated with lowering the risk of cardiovascular disease: _____

    - Transports endogenous triglycerides from the liver to muscle and other tissue _____

    - Principal carrier of cholesterol to the body's tissues _____

    - Apolipoprotein A-1 and A-2 are the major apoproteins found in _____

    - The endogenous pathway involves the metabolism of _____

    - Transports dietary triglycerides from the intestine to the liver _____

Answers

Identify the following as HDL, LDL, VLDL, or Chylomicrons.

- Associated with lowering the risk of cardiovascular disease: <u>HDL</u>

- Transports endogenous triglycerides from the liver to muscle and other tissue: <u>VLDL</u>

- Principal carrier of cholesterol to the body's tissues: <u>LDL</u>

  Apolipoprotein A-1 and A-2 are the major apoproteins found in: <u>HDL</u>

- The endogenous pathway involves the metabolism of: <u>VLDL</u>

- Transports dietary triglycerides from the intestine to the liver: <u>Chylomicrons</u>

CHAPTER 16
Electrolytes and Blood Gases

Clinical Significance
Electrolytes are minerals that are found in body tissues and blood in the form of dissolved salts. As electrically charged particles, electrolytes help move nutrients into and wastes out of the body's cells, maintain a healthy water balance, and help stabilize the body's acid/base (pH) level (1, 24, 29, 33, 48).

Sodium
Most of the body's sodium is found in the fluid outside of the body's cells, where it helps to regulate the amount of water in the body

Potassium Is found mainly inside the body's cells. A small but vital amount of potassium is found in the plasma, the liquid portion of the blood. Potassium plays an important role in regulating muscle contraction. Monitoring potassium is important as small changes in the potassium level can affect the heart's rhythm and ability to contract.

Chloride
Is moves in and out of the cells to help maintain electrical neutrality (concentrations of positively charged cations and negatively charged anions must be equal) and its level usually mirrors that of sodium. Due to its close association with sodium, chloride also helps to regulate the distribution of water in the body

Bicarbonate
The main job of bicarbonate (or total CO_2, an estimate of bicarbonate), which is released and reabsorbed by the kidneys, is to help maintain a stable pH level (acid-base balance) and, secondarily to help maintain electrical neutrality. Bicarbonate also plays an important role in the transport of CO_2: much of the CO_2 produced by the body's tissues is transported in the blood as bicarbonate to the lungs, where it is exhaled

The foods you eat and the fluids you drink provide the sodium, potassium, and chloride your body needs. The kidneys help maintain proper levels by reabsorption or by elimination into the urine. The lungs provide oxygen and regulate CO_2. The CO_2 is produced by the body and is in balance with bicarbonate. The overall balance of these chemicals is an indication of the functional well-being of several basic body functions. They are important in maintaining a wide range of body functions, including heart and skeletal muscle contraction and nerve signaling.

Any disease or condition that affects the amount of fluid in the body, such as dehydration, or affects the lungs, kidneys, metabolism, or breathing has the potential to cause a fluid, electrolyte, or pH imbalance (acidosis or alkalosis). Normal pH must be maintained within a narrow range of 7.35-7.45 and electrolytes must be in balance to ensure the proper functioning of metabolic processes and the delivery of the right amount of oxygen to tissues.

Anion Gap
A related test is the anion gap, which is a value calculated using the results of an electrolyte panel. It reflects the difference between the positively charged ions (called cations) and the negatively charged ions (called anions). An abnormal anion gap is non-specific—it does not diagnose a specific disease or illness—but it can suggest certain kinds of metabolic or respiratory disorders or the presence of toxic substances.

While sodium, potassium, chloride, and bicarbonate are commonly measured together as the electrolyte panel, they can also each be ordered individually for diagnosis/monitoring of conditions that affect specific electrolytes. The body also contains other electrolytes that are not part of the "electrolyte panel" but may also be ordered by your healthcare practitioner. These include calcium (Ca^{2+}), magnesium (Mg^{2+}), and phosphate (PO_4^{3-}).

Acidosis and Alkalosis
Acidosis and alkalosis describe the abnormal conditions that result from an imbalance in the pH of the blood caused by an excess of acid or alkali (base). This imbalance is typically caused by some underlying condition or disease. Normal blood pH must be maintained within a narrow range, typically 7.35-7.45, to ensure the proper functioning of metabolic processes and the delivery of the right amount of oxygen to tissues. Acidosis refers to an excess of acid in the blood that causes the pH to fall below 7.35, and alkalosis refers to an excess of base in the blood that causes the pH to rise above 7.45. Many conditions and diseases can interfere with pH control in the body and cause a person's blood pH to fall outside of healthy limits.

Normal body functions and metabolism generate large quantities of acids that must be neutralized and/or eliminated to maintain blood pH balance. Most of the acid is carbonic acid, which is created from carbon dioxide (CO_2) and water. CO_2 is produced as the body uses glucose (sugar) or fat for energy. Lesser quantities of lactic acid, ketoacids, and other organic acids are also produced.

The lungs and kidneys are the major organs involved in regulating blood pH. The lungs flush acid out of the body by exhaling CO_2. Raising and lowering the respiratory rate alters the amount of CO_2 that is breathed out, and this can affect blood pH within minutes.

The kidneys excrete acids in the urine, and they regulate the concentration of bicarbonate (HCO^{3-}, a base) in blood. Acid-base changes due to increases or decreases in HCO^{3-} concentration occur more slowly than changes in CO_2, taking hours or days. Both of these processes are always at work, and they keep the blood pH in healthy people tightly controlled.

Buffering systems that resist changes in pH also contribute to the regulation of acid and base concentrations. The main buffers in blood are hemoglobin, plasma proteins, CO_2, bicarbonate, and phosphates. The absolute quantities of acids or bases are less important than the balance between the two and its effect on blood pH (49-52).

Acidosis
Occurs when blood pH falls below 7.35. It can be due to increased acid or decreased base:
- Increased acid production within the body
- Consumption of substances that are metabolized to acids
- Decreased acid excretion
- Increased excretion of base

Alkalosis
Occurs when blood pH rises above 7.45. It can be due to decreased acid or increased base:
- Electrolyte disturbances caused by, for example, prolonged vomiting or severe dehydration
- Administration or consumption of base
- Hyperventilation (with increased excretion of acid in the form of CO_2)

Any disease or condition that affects the lungs, kidneys, metabolism or breathing has the potential to cause acidosis or alkalosis. Acid-base disorders are divided into two broad categories:

Respiration
Disorders that affect respiration and cause changes in pH due to changes in CO_2 concentration are called respiratory acidosis (low pH) and respiratory alkalosis (high pH). Respiratory acid-base disorders are commonly due to lung diseases or conditions that affect normal breathing.

Metabolism
Disorders that affect metabolism and cause changes in pH due to either increased acid production or decreased base are called metabolic acidosis (low pH) and metabolic alkalosis (high pH). Metabolic acid-base disorders may be due to kidney disease, electrolyte disturbances, severe vomiting or diarrhea, ingestion of certain drugs and toxins, and diseases that affect normal metabolism (e.g., diabetes).

Signs and Symptoms
Mild acidosis may not cause any symptoms, or it may be associated with nonspecific symptoms such as fatigue, nausea, and vomiting. Acute metabolic acidosis may also cause an increased rate and depth of breathing, confusion, and headaches, and it can lead to seizures, coma, and in some cases death. Symptoms of alkalosis are often due to associated hypokalemia and may include irritability, weakness, and muscle cramping.

Definitions and Key Terms

Anemic hypoxia
Hypoxia is a reduced supply of oxygen to the tissues resulting from a decrease in the amount of hemoglobin or number of erythrocytes in the blood.

Colligative properties
Properties of solutions that depend on the number of particles in the solution; examples include: osmotic pressure, boiling point elevation, freezing point depression, and vapor pressure lowering.

Electrolytes
Charged low-molecular-mass molecules present in plasma and cytosol, usually ions of Sodium, Potassium, Calcium, Magnesium, Chloride, Bicarbonate, Phosphate, Sulfate, lactate.

Electrolyte exclusion effect
Electrolytes are excluded from the fraction of total plasma volume that is occupied by solids, which leads to underestimation of electrolyte concentration by some methods.

Hemoglobin (Hb)
An oxygen-carrying, heme-containing protein abundant in red blood cells.

Ion-selective electrode (ISE)
A type of special-purpose, potentiometric electrode consisting of a membrane selectively permeable to a single ionic species. The potential produced at the membrane–sample solution interface is proportional to the logarithm of the ionic activity or concentration.

Osmolal gap
A difference between the observed and calculated osmolalities in serum analysis.

Osmometry
Technique for measuring the concentration of dissolved solute particles in a solution.

Osmotic pressure
The pressure required to stop osmosis through a semipermeable membrane between a solution and pure solvent.

Oximetry
A technique used to determine the oxygen saturation in arterial blood.

Sweat chloride
The concentration of chloride in sweat; increased sweat chloride is characteristic of cystic fibrosis.

Water homeostasis
The body process that maintains a balance of water intake and output.

Electrolytes and Blood Gases Problems

1. List examples of electrolytes found in plasma water, interstitial fluid, and intracellular water.

2. Identify the analytes required to calculate anion gap and osmolality.

3. State the specific fluid compartments that make up total body water

4. What is the difference between serum and plasma.

5. List five examples of body fluids that are assayed for electrolyte composition.

6. Identify four methods used to measure chloride in sweat.

7. Define anion gap and discuss its clinical significance.

8. Describe an appropriate course of action for the measurement of sodium ion in a sample that shows gross lipemia.

9. Describe the homeostatic regulation of sodium, potassium, and chloride in body water compartments.

10. What is the most probable acid/base imbalance in the patient described below, respiratory alkalosis or acidosis?

    26-year-old male presents with the following signs and symptoms:
    Extreme hysteria
    Hyperventilating
    Pulse 110 BPM
    Lab data:
    pH = 7.550
    pCO_2 = 27 mmHg
    HCO_3^- = 27 mEq/L
    pO_2 = 98 mmHg
    Answer: Respiratory alkalosis

11. Which condition, renal failure or hypoaldosteronism, will cause an increase anion gap?
    Answer: Renal failure

12. The anion gap is determined from which of the following groups of electrolytes?
    a. Sodium, chloride, potassium, and calcium
    b. Sodium, chloride, potassium, and HCO_3^-
    c. Sodium, chloride, potassium, and phosphorus
    d. $TeCO_2$, chloride, potassium, and magnesium
    e. Answer: b

13. List the conditions that may be characterized by an increased anion gap?
    a. Salicylate intoxication
    b. Diabetes mellitus
    c. Lactate acidosis
    d. All of the above
    Answer: d

14. A sweat chloride greater than 60 mEq/L is consistent with what disorder?
    Answer: Cystic fibrosis

15. Symptoms including metal confusion, weakness, tingling sensations, respiratory muscle weakness are all associated with which of the following disorders?
    a. Hyperkalemia
    b. Hypokalemia
    c. Hyperphosphatemia
    d. Hypochloremia
    Answer: a

16. Patients with hyponatremia (low sodium) may show signs of hypovolemia which is defined as:
    a. Low blood protein levels.
    b. High blood potassium levels.
    c. Low water volume.
    d. Low blood pressure.
    Answer: c

17. A 43-year-old female arrived at the emergency department with altered level of consciousness. She complained of chronic but moderate severity acid indigestion. This was partially relieved by taking antacids. Her blood gases results are as follows:

    RESULTS:
    pH = 7.60
    pCO_2 = 38 mm Hg
    HCO_3^- = 40 mEq/L

    This patient demonstrates:
    a. Respiratory acidosis.
    b. Metabolic alkalosis.
    c. Respiratory alkalosis.
    d. Metabolic acidosis.
    Answer: b

Index

5'-Nucleotodase, 69
abnormal, 60, 61, 72, 85, 90
Absorbance, 1, 4
Absorptivity, 1
Acidosis, 84, 90
Acute pancreatitis, 73, 76
Affinity, 10, 13, 21, 23
Alanine Aminotransferase, 68
Albumin, 50, 52, 54, 57, 80
Alkaline Phosphatase, 69
Alkalosis, 90, 91
Alkaptonuria, 51, 58
Aminoacidopathies, 51
Ammonia, 62
Amperometry, 7
Ampholyte, 18
Amyloid, 53
Analytical specificity, 25
Antibody, 21, 23
Antigen, 21, 23
Apolipoproteins, 86
Applied pharmacokinetics, 42
Argininosuccinic Aciduria, 51
Aspartate Aminotransferase, 68, 71
Asymmetric PCR, 38
Atherosclerosis, 86
Atomic absorption spectrophotometry, 2
Avidity, 21
Azotemia, 59, 60, 67
Base pair, 32
Base peak, 15, 17
Beers law, 2
Bence-Jones proteins, 55
Bias, 25, 30
Bilirubin, 70, 76
Bioavailability, 42, 45
Biosensor, 7
Biuret Assay, 54
Brain Natriuretic Peptide, 53, 57
Calibration, 25
Cardiac Troponin, 53
Centromere, 32
Chemiluminescence, 2, 6
chromatogram, 15, 16, 17
chromatography, 10, 11, 12, 13, 14, 15, 52
Chromosome, 32
Chylomicrons, 85, 88
Citrullinemia, 51
Coefficient of variation, 25
Copy number variant, 33
Coulometry, 7

Creatine, 60, 61, 70, 71, 73, 75
Creatinine, 59, 60, 61, 63, 64, 67
Cystatin C, 53
Cystinuria, 51
Densitometry, 18
DeRitis, 69, 75
Diabetes, 77, 78, 79, 81, 94
Difference plot, 26
DNA methylation, 33
DNA microarray, 38
Dose, 42, 43
Drug half-life, 43
Dye Binding, 54
Electrochemical cell, 7
Electrochemiluminescence, 2
Electrode, 7, 8
Electrolytes, 89, 92, 93
Electropherogram, 18
electrophoresis, 18, 19, 20, 21, 22, 37, 40, 52, 55
Electrophoresis, 18, 20, 55
Endosmosis, 18
Epigenetics, 33
Fibronectin, 53
First-pass effect, 43
Flow cytometry, 2, 38
Fluorescence, 2, 6, 38
Folin–Lowry Method, 54
Fragment ion, 15, 17
Gamma Glutamyl Transferase, 69
Genome, 33
Globulin, 52, 54
Glucagon, 81, 83
Glucose, 78, 79, 80, 81, 82
Half-life, 42
Hapten, 21
Hemoglobin, 80, 92
Histones, 34
Homocystinuria, 51
Hook effect, 22
Hyperglycemia, 77, 84
Hypoglycemia, 77, 79, 81, 84
immunoassay, 6, 21, 22, 23, 24
Immunoassay, 22
ionization, 15, 16, 17
Ionization, 15
Ion-selective electrode, 8, 92
Isovaleric Acidemia, 51
Jaffe reaction, 64
Kidney, 57, 62, 64, 69
Kjedahl, 54
Kwashiorkor, 55

Lactate Dehydrogenase, 70, 71
Lesch-Nyhan Syndrome, 67
Levey-Jennings chart, 26
lipoproteins, 85, 86
Liver, 52, 57, 68, 69, 70, 82
Maple Syrup Urine Disease, 51, 58
Mass spectrometer, 15
Mean, 26, 31
Median, 27
Metabolic, 14, 91, 94
Metabolism, 47, 91
Myoglobin, 53, 57, 70, 72
Nephelometry, 2
Nephron, 62
Nonprotein Nitrogen Compounds, 59
Odds ratio, 27
Ordinary least-squares regression, 27
Osmometry, 92
Oximetry, 92
Pharmacodynamics, 43
Pharmacokinetics, 43
Phenotype, 34
Phenylketonuria, 51, 58
Photometry, 1
Phred score, 35
Polymerase chain reaction, 39
Potentiometry, 8
Pseudogene, 35
Pyruvate, 81
Regression analysis, 27
Sothern blotting, 22
Spectrophotometry, 3
Standard deviation, 27
Student t distribution, 28
Systematic error, 28
Therapeutic drug monitoring, 14, 42, 43
Toxicology, 43
Traceability, 28
Transmittance, 3, 49
Troponins, 71
Turbidimetry, 3, 6
Tyrosinemia, 51, 58
Uncertainty, 28
Urea, 59, 64, 67
Uric Acid, 61
Volume of Distribution, 42
Western blotting, 22, 23
Westgard Rules, 28
Wick flow, 19
Xenobiotics, 14, 44

Reference

1. Matysik F-M. *Bioanalytical Reviews,*. Cham: Springer International Publishing: Imprint: Springer, 2014:1 online resource (VII, 264 pages 162 illustrations, 22 illustrations in color.)
2. Chung HJ, Sulkin MS, Kim JS, Goudeseune C, Chao HY, Song JW, et al. Stretchable, multiplexed pH sensors with demonstrations on rabbit and human hearts undergoing ischemia. *Adv Healthc Mater.* 2014;3(1):59-68.
3. Valencia, California: American Scientific Publishers; 2013.
4. Cappiello A. *Advances in LC-MS instrumentation.* Amsterdam ; Boston, MA: Elsevier; 2007.
5. Almirall J, Akmeemana A, Lambert K, Jiang P, Bakowska E, Corzo R, et al. Determination of seventeen major and trace elements in new float glass standards for use in forensic comparisons using laser ablation inductively coupled plasma mass spectrometry. *Spectrochim Acta Part B At Spectrosc.* 2021;179.
6. Keren DF. *Protein electrophoresis in clinical diagnosis.* Chicago, IL: American Society for Clinical Pathology Press; 2012.
7. Keren DF. *High-resolution electrophoresis and immunofixation : techniques and interpretation.* Boston: Butterworth-Heinemann; 1994.
8. Keren DF. *High-resolution electrophoresis and immunofixation : techniques and interpretation.* Boston: Butterworths; 1987.
9. Wolf PL, Griffiths JC, and Koett J. *Interpretation of electrophoretic patterns of proteins and isoenzymes : a clinical pathologic guide.* New York: MASSON Pub. USA; 1981.
10. Shekhar HU. *Biochemistry and molecular biology in the post genomic era.* New York: Nova Science Publishers, Inc.,; 2019:1 online resource.
11. Wilson K, and Walker JM. *Principles and techniques of biochemistry and molecular biology.* Cambridge, UK New York: Cambridge University Press; 2009.
12. Jain KK. *Applications of biotechnology in oncology.* New York: Humana Press; 2014.
13. Barciszewski J, Erdmann VA, and Jurga S. *RNA Technologies,*. Cham: Springer International Publishing : Imprint: Springer,; 2015:1 online resource (X, 349 pages 124 illustrations, 78 illustrations in color.
14. Członkowska A, and Schilsky ML. *Wilson disease.* Amsterdam, Netherlands: Elsevier; 2017.
15. Rifai N, Horvath AR, and Wittwer C. *Principles and applications of molecular diagnostics.* Amsterdam, Netherlands ; Oxford, United Kingdom ; Cambridge, MA: Elsevier; 2018.
16. Nag S. *The blood-brain and other neural barriers : reviews and protocols.* New York, N.Y.: Humana Press; 2011.
17. Tang Y-W, and Stratton CW. Cham: Springer International Publishing : Imprint: Springer,; 2018:1 online resource (XIV, 541 pages 94 illustrations, 67 illustrations in color.
18. Lacey LA. *Manual of techniques in invertebrate pathology.* Oxford ; New York: Academic Press imprint of Elsevier Science; 2012.
19. Davies PDO, Gordon SB, and Davies G. *Clinical tuberculosis.* Boca Raton: CRC Press, Taylor & Francis Group; 2014.
20. Filloux A, Ramos J-L, and Springer Science+Business Media. *Pseudomonas : methods and protocols.* New York: Humana Press; 2014.
21. Provan D, and Gribben J. *Molecular hematology.* Hoboken, NJ: Wiley-Blackwell; 2020.
22. Provan D, and Gribben J. Hoboken, NJ: Wiley-Blackwell,; 2020:1 online resource.
23. Moyer TP, and Boeckx RL. *Applied therapeutic drug monitoring.* Washington, DC: American Association for Clinical Chemistry; 1984.
24. Baranowska I. Cham: Springer International Publishing : Imprint: Springer,; 2016:1 online resource (XI, 453 pages 76 illustrations, 11 illustrations in color.
25. Passi GR, Wakchaure A, and Jaiswal SP. Clinical and Genetic Spectrum of 50 Children with Inborn Errors of Metabolism from Central India. *Indian J Pediatr.* 2021.
26. Turksen K. *Stem Cell Biology and Regenerative Medicine,*. Cham: Springer International Publishing : Imprint: Humana,; 2018:1 online resource (XI, 177 pages 28 illustrations, 26 illustrations in color.
27. Burtis CA, Bruns DE, Sawyer BG, and Tietz NW. *Tietz fundamentals of clinical chemistry and molecular diagnostics.* St. Louis, Missouri: Elsevier/Saunders; 2015.
28. Taylor EH. *Clinical chemistry.* New York: Wiley; 1989.
29. Kaplan A, and Kaplan A. *Clinical chemistry : interpretation and techniques.* Baltimore: Williams & Wilkins; 1995.
30. Arneson W, and Brickell J. *Clinical chemistry : a laboratory perspective.* Philadelphia: F.A. Davis Co.; 2007.
31. Bertholf RL, and Winecker RE. *Chromatographic methods in clinical chemistry and toxicology.* Chichester, England ; Hoboken, NJ: John Wiley & Sons; 2007.

32. Caroli S, and Záray G. *Analytical techniques for clinical chemistry : methods and applications.* Hoboken: Wiley; 2012.
33. Larson D, Hayden J, and Nair H. *Clinical chemistry : fundamentals and laboratory techniques.* St. Louis, Missouri: Elsevier/Saunders; 2017.
34. Cappiello A, and Palma P. *Advances in the use of liquid chromatography mass spectrometry (LC-MS) : instrumentation developments and application.* Amsterdam, Netherlands: Elsevier; 2018.
35. Palmer T, and Bonner PLR. *Enzymes : biochemistry, biotechnology and clinical chemistry.* Chichester: Horwood; 2007.
36. . New York ,: Wiley-Interscience; 1960:11 volumes.
37. Harnisch LO, Baumann S, Mihaylov D, Kiehntopf M, Bauer M, Moerer O, et al. Biomarkers of Cholestasis and Liver Injury in the Early Phase of Acute Respiratory Distress Syndrome and Their Pathophysiological Value. *Diagnostics (Basel).* 2021;11(12).
38. Novotny JF, and Sedlacek F. *Bilirubin : chemistry, regulation, and disorder.* Hauppauge, N.Y.: Nova Science Publishers; 2012.
39. Pinto PC, Ronnau C, Burchardt M, and Wolff I. Kidney Cancer and Chronic Kidney Disease: Too Close for Comfort. *Biomedicines.* 2021;9(12).
40. Stakhneva EM, Striukova EV, and Ragino YI. Proteomic Studies of Blood and Vascular Wall in Atherosclerosis. *Int J Mol Sci.* 2021;22(24).
41. Jia W. Singapore: Springer Singapore : Imprint: Springer,; 2018:1 online resource (XVI, 215 pages 128 illustrations, 13 illustrations in color.
42. Bast P. *Biochemistry and molecular biology in the post genomic era.* New York: Nova Science Publishers, Inc.,; 2020:1 online resource.
43. Bast P. *Advanced glycation end-products : sources and effects.* New York: Nova Science Publishers, Inc.; 2020.
44. White JR, and American Diabetes Association. *2020-21 guide to medications for the treatment of diabetes mellitus.* Arlington, VA: American Diabetes Association; 2020.
45. Vella A. *Clinical dilemmas in diabetes.* Hoboken, NJ: Wiley-Blackwell; 2021.
46. *Biochemistry of lipids, lipoproteins and membranes.* Amsterdam ; Boston: Elsevier; 2016.
47. Wilson P. *Atlas of atherosclerosis : risk factors and treatment.* Philadelphia: Current Medicine; 2000.
48. Bastidas DM, and Cano E. Cham: Springer International Publishing : Imprint: Springer,; 2018:1 online resource (XI, 105 pages).
49. Matysik F-M. *Bioanalytical Reviews,.* Cham: Springer International Publishing : Imprint: Springer,; 2017:1 online resource (VII, 344 pages 70 illustrations, 58 illustrations in color.
50. Nurse CA, and International Society of Arterial Chemoreception. *Arterial chemoreception : from molecules to systems.* Dordrecht ; New York: Springer Verlag; 2012.
51. Kacmarek RM, Hess D, and Stoller JK. *Monitoring in respiratory care.* St. Louis: Mosby; 1993.
52. Shapiro BA, and Shapiro BA. *Clinical application of blood gases.* Chicago: Year Book Medical Publishers; 1989.

www.ingramcontent.com/pod-product-compliance
Lightning Source LLC
Chambersburg PA
CBHW082212220526

45470CB00010B/3131